Fleet Manager's Guide to Vehicle Specification and Procurement

Second Edition

Other SAE titles of interest:

Alternative Cars in the 21st Century
A New Personal Transportation Paradigm
(Second Edition)
By Robert Q. Riley
(Order No. R-227)

Turnaround: How Carlos Ghosn Rescued Nissan
By David Magee
(Order No. B-860)

Utility Vehicle Design Handbook
(Second Edition)
Edited by John F. Hoelzle, O.C. Amrhyn, and
Gary A. McAlexander
(Order No. AE-16)

For more information or to order a book, contact SAE at
400 Commonwealth Drive, Warrendale, PA 15096-0001;
phone (724) 776-4970; fax (724) 776-0790;
e-mail CustomerService@sae.org;
website http://store.sae.org.

Fleet Manager's Guide to Vehicle Specification and Procurement

Second Edition

John E. Dolce

Warrendale, Pa.

For permission and licensing requests, contact:

SAE Permissions
400 Commonwealth Drive
Warrendale, PA 15096-0001 USA
E-mail: permissions@sae.org
Tel: 724-772-4028
Fax: 724-772-4891

Library of Congress Cataloging-in-Publication Data

Dolce, John.
   Fleet manager's guide to vehicle specification and procurement /
John E. Dolce. --2nd ed.
     p.   cm.
   Includes index.
   ISBN 0-7680-0981-2
   1. Motor vehicle fleets--Management. 2. Motor vehicles--Specifica-
tions. 3. Motor vehicles--Purchasing. I. Title.

TL165.D65 2003
388.3'2'068--dc21

                                                                    2003054446

SAE
400 Commonwealth Drive
Warrendale, PA 15096-0001 USA
E-mail: CustomerService@sae.org
Tel: 877-606-7323 (inside USA and Canada)
            724-776-4970 (outside USA)
Fax: 724-776-1615

Copyright © 1992, 2003   SAE International
ISBN 0-7680-0981-2
SAE Order No. R-332
Printed in the United States of America.

This book is dedicated to my parents, Stella and John Dolce,
and to my wife, Arlene, and my children,
Larraine, Madeline, Craig, and Stephanie.

Acknowledgments

A special thanks to my wife, Arlene, and family, Larraine, Madeline, Craig, and Stephanie, for their support, and to Laura Zabriskie for her help in organizing and typing the first edition of this book, and for her participation in supporting these efforts.

None of this would have been possible without the people of the transportation industry. Their dedication and desire to perform is what makes being a part of this industry enjoyable. Fleet people have a wealth of knowledge, and their efforts are supporting significant improvements in vehicle performance.

I hope this work will bring greater understanding of transportation industry management and, in some small way, contribute to its improvement.

Preface

This book is written for all supervisors and managers with a responsibility in the areas of vehicle specification and procurement. Practical, cost-effective principles and practices will be defined and illustrated.

It will serve as a field reference to practices and procedures used in the daily work operations, with emphasis on the implementation of these cost-effective principles. Important ingredients of applying these practices and procedures are fact-finding, analysis, alternative selection, and communication to ensure the proper application of these principles.

The book is written in simple, basic terms using the profit dollar as the basis for guidelines of actions to be taken. These guidelines are generally accepted by the commercial motor transportation industry. The action taken can be referred to by all levels of management because the common reference point is the profit and growth dollar.

Productivity savings will be realized. Each management person involved in the fleet management area should be able to understand this information and apply it to realize a productivity savings.

Change is necessary for a productivity improvement. Change costs money. Will the cost of change be recovered in the savings resulting from it? If so, it is worth implementing.

We are moving toward a proactive approach rather than reactive, and this work as part of the mission supports that vision.

The applications and implementation of these principles in a stress environment takes skill. It is these principles and skills that are the subject matter of this text.

It is hoped that this work will increase fleet managers' understanding of the transportation industry and serve as a guide in adding to their skills so, when applied, the results will extend to improvements in the performance of their fleets.

Table of Contents

Chapter I

General Introduction and Overview of Vehicle Specification and Procurement

General

The process of purchasing a light, medium, heavy, or on-/off-road vehicle is a detailed activity. It must be defined, prioritized, sequenced, and performed annually. This will ensure that a workable vehicle is delivered to be put into service in a trouble-free condition and used efficiently for its intended life.

Understanding Productivity

Productivity is a balance and combination of several meaningful items.

1. Capital investment: old facilities and equipment are a handicap
2. Formal employee involvement: quality circle; and buy-in
3. Quality control: third-party involvement; focus on the process
4. Measure outputs: measure change due to corrective action
5. Work methods + work knowledge + work habits = training
6. Direct-indirect labor definitions = staffing, balance, and mix
7. Resistance to change: politics, emotion, disagreement, misunderstanding
8. Fact-finding education; motivation; leadership; common goals
9. Money to finance to measure: common goals
10. Attitude: management; supervision; worker; 80/20 rule
11. Incentive programs; values; commitment

12. Recognition: wages versus company ownership
13. One-shot savings; maintain new level; sustain gains; new ideas

Vehicle Replacement Policy

Most, if not all, fleets have a vehicle replacement policy. The policies are defined in terms of miles, age, or both. These policies have evolved through experience, instinct, cost analysis, and/or emotion.

These policies provide a basis for an annual review of a fleet to identify the vehicles that meet these criteria. Annually, the vehicles for replacement would be listed in a descending order, based on their age or mileage. Their age would be forecasted in the coming replacement year based on their anticipated delivery date. If a vehicle is to be replaced at 10 years old or 100,000 miles, it would be forecasted in the delivery year when it would meet this criterion. At January of the present year, a vehicle has 90,000 miles and is nine years old. The next budget year starts in January. This vehicle travels 5,000 miles per year. In 18 months, or June of the coming year, it will be qualified for replacement at 10 years of age with 100,000 miles on it. The delivery of the new vehicle would be targeted for December of that year and, of course, budgeted for its arrival at that time. Construction equipment such as backhoes, bulldozers, graders, bucket loaders, and warehouse equipment would use time, fuel use, and/or hour-meters as criteria for replacement. Five thousand hours seems to be a targeted evaluation level for replacement evaluation.

To ensure that this vehicle is being replaced efficiently, the vehicle would be inspected in the presence of the user, the maintenance staff, and the transportation staff that is directing the replacement process. With all the direct personnel involved with the replacement process present, a discussion can take place evaluating the pros and cons so that at the end of this meeting, each of the group has agreed to the course of action and their expectations. The vehicle assessment report, as shown in Figures 1-1 and 1-2, can quantify and sequence vehicle conditions and can be adjusted efficiently for variable capital resources.

Whether to complete the replacement or extend vehicle life, to upgrade or downsize, to add to or delete some components, to measure wants versus needs, and to agree on a course of action are of utmost importance. This provides for a start of the vehicle specification and replacement process.

VEHICLE NO.:		VEHICLE MAKE:		MODEL:		YEAR:		
Rating Legend:	5 = Excellent	4 = Very Good		3 = Good	2 = Average		1 = Poor	
			Inspection Results					
Description	Awarded Rate	Multiplier		Actual Rate	Max. Rate		Factor - Score	
Section 01 Body/Interior		TOTAL MAX. POINTS 15						10.8
Rust	3	× 20	=	60	100			
Condition	3	× 20	=	60	100			
Accident Damage	4	× 40	=	160	200			
Glass	5	× 10	=	50	50			
General Interior Condition	5	× 10	=	30	50	360	72%	
			Total	360	500	500		
Section 02 Tires		TOTAL MAX. POINTS 15						9.0
Tread Wear	3	× 60	=	180	300			
Sidewall Cracking	3	× 40	=	120	200	300	60%	
			Total	300	500	500		
Section 03 Driveability		TOTAL MAX. POINTS 15						9.0
Road Test	3	× 100	=	300	500	300	60%	
			Total	300	500	500		
Section 04 Brake System		TOTAL MAX. POINTS 15						15.0
Stop Ability	5	× 60	=	300	300			
Emergency	5	× 40	=	200	200	500	100%	
			Total	500	500	500		
Section 05 Steering/Suspension		TOTAL MAX. POINTS 10						8.0
Loose	4	× 20	=	80	100			
Vibration	4	× 20	=	80	100			
Pulling	4	× 20	=	80	100			
Standing Appearance (level/square to road)	4	× 40	=	160	200	400	80%	
			Total	400	500	500		
Section 06 Electrical		TOTAL MAX. POINTS 10						10.0
Once Over Safety Items	5	× 100	=	500	500	500	100%	
			Total	500	500	500		
Section 07 Driveline		TOTAL MAX. POINTS 10						10.0
Noise	5	× 20	=	100	100			
Vibration	5	× 20	=	100	100			
Leaks	5	× 20	=	100	100			
Shifting	5	× 40	=	200	200	500	100%	
			Total	500	500	500		
Section 08 Engine		TOTAL MAX. POINTS 10						8.0
Leaks	4	× 30	=	120	150			
Rough Running	4	× 40	=	160	200			
Noise	4	× 30	=	120	150	400	80%	
			Total	400	500	500		
	TOTAL 92		TOTAL RATE AWARDED	3260		TOTAL SCORE	79.80	

Total Score Legend:
3200–4000 Excellent
2400–3200 Very Good
1600–2400 Good
800–1600 Average
0–800 Poor

Awarded Score Legend:
88–110 Excellent
66–88 Very Good
44–66 Good
22–44 Average
0–22 Poor

Figure 1-1
Vehicle Assessment Report
Light-Duty Vehicle

VEHICLE NO:		VEHICLE MAKE:			MODEL:		YEAR:	
Rating Legend:	5 = Excellent		4 = Very Good		3 = Good	2 = Average	1 = Poor	
	Awarded				Actual	Max.		
Description	Rate		Multiplier		Rate	Rate	Factor-Score	
Section 01 Chassis			TOTAL MAX. POINTS = 15					10.8
Rust and Corrosion	3	×	20	=	60	100		
Condition	3	×	20	=	60	100		
Accident Damage	4	×	40	=	160	200		
Glass	5	×	10	=	50	50		
Interior	3	×	10	=	30	50	360	72%
			Total		360	500	500	
Section 02 Tires			TOTAL MAX. POINTS = 15					9.0
Tread Wear	3	×	60	=	180	300		
Sidewall condition	3	×	40	=	120	200	300	60%
			Total		300	500	500	
Section 03 Body Mounted Equipmt.			TOTAL MAX. POINTS = 15					9.0
Dump Body Exterior	3	×	20	=	60	100		
Dump Body Interior	3	×	20	=	60	100		
Tail Gate	3	×	20	=	60	100		
Extension Boards	3	×	20	=	60	100		
Lift Cylinder	3	×	20	=	60	100	300	60%
			Total		300	500	500	
Section 04 Brake System			TOTAL MAX. POINTS = 15					12.0
Service Brakes	4	×	60	=	240	300		
Emergency Brakes	4	×	40	=	160	200	400	80%
			Total		400	500	500	
Section 05 Steering/Suspension			TOTAL MAX. POINTS = 15					12.0
Looseness	4	×	20	=	80	100		
Vibration	4	×	20	=	80	100		
Pulling	4	×	20	=	80	100		
Parallel to Ground	4	×	40	=	160	200	400	80%
			Total		400	500	500	
Section 06 Engine and Driveline			TOTAL MAX. POINTS = 25					16.0
Leaks	2	×	20	=	40	100		
Vibration	2	×	20	=	40	100		
Noise	3	×	20	=	60	100		
Shifting	4	×	20	=	80	100		
Rough Running	5	×	20	=	100	100	320	64%
			Total		320	500	500	
	TOTAL 79		TOTAL RATE AWARDED		2080	TOTAL SCORE		68.8

Total Score Legend:
2400–3000 Excellent
1800–2400 Very Good
1200–1800 Good
600–1200 Average
0–600 Poor

Awarded Score Legend:
92–115 Excellent
69–92 Very Good
46–69 Good
23–46 Average
0–23 Poor

Figure 1-2
Vehicle Assessment Report
Heavy-Duty Vehicle

Of course, replacement of a vehicle can be accelerated by the user and maintenance staff due to deteriorated vehicle condition, accident damage, and/or operating environment changes when one-time repair or cumulative maintenance costs accumulate to 30% of the residual value of the vehicle. This is the time to initiate the discussion.

With cars and light trucks (up to 10,000 lb GVW), a physical inspection is not necessary in evaluating whether the vehicle should be replaced. Cost information, utilization information along with fleet mix, and density data are enough to plan a replacement process (Figure 1-1).

Vehicles 10,001 lb GVW and higher should be physically inspected to look for wear and tear that would show the need for modifications to the replacement vehicles. For example, many dented grills might denote the need for vehicle front guards, broken lights might denote the need for recessed lights, worn body shelves might require better materials, and rust might determine the need for improved materials and/or metal processing requirements (Figure 1-2).

The physical inspection process should include pictures of the vehicle, notes on alternatives discussed, and action agreed to be taken as of that date by the people assembled.

Of course, all that is subject to change as the process continues and the overall strategies of the company are considered. Money might be held back, work methods might be changing, and if the preceding documentation is available, it can be referred to for clarifying needs.

Planning Your Budget

Our present year budget has come as a result of two years of defining, explaining, strategizing, calculating, balancing, and cooperating. Hopefully, you have created it with the goal of helping support a competitive posture so you can cost-effectively support your internal and external customers. Good budgets not only meet needs, but also remain flexible to changing demands.

So, let us take a look at what goes into determining a budget.

We know the external customers' needs: more service, that is also reliable, for less cost and immediate response to unscheduled events such as storms, brownouts, and population shifts.

Your internal customers have the following resources available to meet those expectations: equipment, people, facilities, supplies, and components.

To satisfy customer needs, work methods, past practices, efficiencies, inefficiencies, and new technologies must be homogenized into present practices.

No matter what level of management (supervisor, manager, etc.), there must be a perception and perspective that is based on cooperation to meet customer expectations efficiently. This sounds simple, but it is not.

The key is to "cooperate internally to be competitive externally." Our business as fleet managers is technical, busy, and complex, but not complicated. A lack of internal cooperation makes it complicated.

That is why the budget process must be a two-year cycle.

Let us look at executive management's questions and concerns when considering our budget: "Where will our company be one, two, five, and even ten years from now? What is our company's industrial, commercial, and residential mix of business going to be? How will the transmission, distribution, generation, and operating resources be positioned to meet the needs of our customers? What kind of contingency plan should we establish for catastrophic events?"

Each department should adjust its resources to meet these challenges with more efficiency. In other words, do more with less. While you might be able to trim some of the fat off your budget, do not cut off some of the "meat" simply to lower the budget cost from the previous year.

Internally, in each department, you probably have a wide variety of personnel of varying experiences, perceptions, and perspectives that, when homogenized, should work in harmony to meet the "cooperate to be competitive" mentality.

If this happens, the company gets better; if not, the company will struggle and instead will become internally competitive. This will mean a reduced efficiency in costs which, in turn, increases the costs to provide services to the external customers.

As a result, the strategy for upper management to relieve the internal tension of the company will be to squeeze the budget, forcing the departments to be more efficient by giving them less capital and less operating money.

"Yes, the beatings will stop when the company is more efficient!"

There should be conversation with each of your internal fleet customers two years prior to the presentation of a capital and operating budget. One year will be needed to explore and analyze what fleet technology is out there, and the second year is for determining the costs because we all know prices do not remain constant.

Things change. Before you put a cost on your needs, it is best to prepare some options:

1. If you need a specific capital outlay for *status quo*, determine what the resultant fleet operating cost will be for staffing, parts, and space.

2. If corporate denies the specified capital requested, determine the resultant operating dollars needed to cover increased costs of aged equipment (factoring in the decision of adding more staff, space, and parts or adding more outsourced revenue). Rule of thumb: Each capital dollar denied adds 30 cents to the operating budget.

3. Invest more capital into work method improvements. This could result in lower operating costs.

It is your responsibility as a fleet manager to be proactive rather than reactive. We must give executive management economic choices so it can fit one of them into a strategic company plan. Having savvy will allow you to cost out your internal clients' needs (not wants) and to cost them out with efficient alternatives for cooperative planning.

In short, plan your work and work your plan. The plan is the budget. You need two years to be able to qualify your needs.

It will take time to educate, guide, and sell the internal customer on what resources are out there.

Budgeting

The replacement process starts with the budgeting process, the putting aside of funds to pay for the vehicle when it is delivered. The company has a strategic plan. This plan includes growth, *status quo*, or consolidation of its operations. The vehicle is a tool to support the operating plan of the company. If cash or capital is not available, in-house vehicle conditions illustrated in the preceding documentation would support the need for borrowing funds or transferring funds to meet the capital needs or the leasing option to maximize cash flow rather than pursue the straight capital purchase. If success is limited in this effort, then discussions could be initiated to rebuild or refurbish the present fleet, extending replacement cycles to a future date when economic conditions improve.

Communicating with Numbers

Language is arbitrary. Words have meanings, and people have meanings for words, which makes efficient communication difficult. Because words tend to be subjective and emotional, let us transpose communication into an objective format. Take the emotion out of the exchange of information, and level the learning process.

The participants and their ability to be patient, of course, control this process. They ask for clarifications and have the courage and energy to endure, which is limited by the frustrations of the exchange of information.

Let us use fuel mileage as a platform. To know miles per gallon and/or gallons per hour, we divide gallons into miles and obtain the total miles per gallon (mpg). We have a base provided by the manufacturer, who states that if you drive this speed with this engine and these chips in the electronic control unit with this gearing and these times, your miles per gallon around the inner city is 5 mpg. In a suburban setting, it is 7 mpg, and over the road, it is 10 mpg.

When we do the calculations in an application, we have a targeted number at which to aim, compliments of the manufacturer. Should we meet that number, the circle of evidence is complete.

If we do not meet that number, we check the accuracy of the elements such as miles and gallons to validate the end result. This points out to us the efficiency of the process. However, the efficiency can be hindered because of human error such as misread numbers, handwriting smudges, numbers written in the wrong places, and scanning irregularities.

Today, this information is in the electronic control unit. Such information is retrievable through wire and wireless efforts, with additional support data such as combustion temperature, oil, water, and air-fuel temperatures. Thus, we can periodically monitor and take action before symptoms appear and preventable failures on unscheduled activity take place.

Using the numbers, we have an objective format and a clear communication for measuring vehicle productivity and uncovering variances in our numbers.

The numbers certainly beat the words in saving time and getting straight to the issue, allowing fleet people to do what they do best: fix the problem proactively. Ten years ago, before fleet electronics were common, we were rewarded for quick and thorough reaction. Since electronics on vehicles and data processing has become a reality, there has been a change for the better to improve performance, which in turn, reduces costs. Today, we can do more with less resources to become competitive, but only if we take advantage of change which, in this case, is electronics and data processing technologies.

More numbers help us. Our vehicle maintenance management systems give us scheduled and unscheduled information, such as preventative maintenance inspection and generated repair costs. We can use this information to look at our road calls and breakdowns as a means to compare corporate costs.

The costs of these areas allow us to focus because the numbers are prioritized. We accept the large figures and devote our attention to the smaller numbers to ask questions, pursue an audit trail, validate our findings, and initiate corrective actions that work. You also must be accurate when making those corrective actions. Today, it is one strike and you are out when you decide to swing. Four balls do not walk, and if you do not swing at a strike over the plate, you are out.

The fleet management process includes many levels of people, and our line with these people is our budgets, including capital and operating areas. That is where the numbers count.

Our management information systems point out activities that necessitate budgeting explanations for actuals that are either too high or too low. Accounting and finance numbers are zero-based, but vehicle management information systems are not designed to be that way.

What is needed to identify issues is our vehicle management information system. Then, when we investigate and resolve these issues, we move into general ledger numbers, payroll, accounts payable, and accounts receivable.

In fact, we must take our action items that were effective through our vehicle management information system and cross into the budget area. It is a cooperative effort. Numbers are objective. Effort levels the playing field to enable us to proceed together with minimum distractions and to identify the problems, not symptoms, and solve them. Cooperate internally to be competitive externally.

Vehicle Inventory

Historical information is used to project future activity. Figure 1-3 shows the number of vehicles Fleet A has inventoried to support its operating plan. The information is broken down by description, size, and model of vehicle; class, internal classification of vehicles; and weight class, an internal delineation of their weights needed to quantify costing and billing information and age. In this case, January of each year is targeted to summarize and to track vehicle inventory counts.

Vehicle Sizes

Figure 1-4 identifies the weight classes of vehicles agreed on by the manufacturer and users for communication on construction and operating costs. The industry collects and references costs to these classes for their measurements to help the industry manage costs efficiently.

What Does Your Fleet Cost?

How much does it cost to own and operate your fleet? Do we look at the budget and operating and capital costs divided by miles and hours? Do we look to our vehicle class costs, where budgeted operating and capital costs are cleared and charged to classes of vehicles delineated by accounting codes? Or do we go to the management information systems we use to obtain cost information by miles, hour, class, and in-depth indices such as warranty, components, parts, labor (direct and indirect), fully burdened labor rate, and more?

The answer is all of the above. Not only must we have accuracy in each category, but each cost system also should define and say the same thing. The purpose of an accounting cost category system is to regulate systems and

DESCRIPTION	CLASS	1994	1995	1996	1997	1998	1999	2000	2001	2002
Weight Class 1										
Executive	01	16	17	14	14	12	12	12	10	10
Intermediate	03	18	8	5	12	23	23	23	23	22
Compact Sedan	04	342	293	344	306	297	231	182	180	178
Compact Sta. Wagon	07	24	19	29	26	30	67	80	79	79
SUBTOTAL		400	337	392	358	362	333	297	292	289
Weight Class 2										
Van Compact	010	0	0	2	6	6	29	36	36	36
Utility Veh. 4x4	012	7	11	9	9	17	19	20	20	20
P/U 4x2 Compact	013	11	45	146	242	241	292	287	279	278
Com. Wagon 4x2	015	4	6	6	7	7	7	7	7	7
Com. Wagon 4x4	016	3	2	7	7	9	9	12	5	5
Van Small	017	129	136	140	139	14	135	125	126	126
Pick-Up 4/2	018	281	258	159	99	92	62	62	62	62
Utility 4x4 Com.	022	0	0	0	0	0	1	1	8	8
SUBTOTAL		435	458	469	509	515	554	550	543	542
Weight Class 3										
Pick-Up 4x4	019	15	15	17	36	39	42	41	42	42
Van Walk-In	021	3	5	8	11	18	22	22	24	24
Truck Light	023	22	21	25	26	24	26	25	25	25
Truck Medium	025	14	14	20	19	13	11	10	10	10
Aerial Bucket Sm.	032	34	48	52	74	72	68	68	67	67
SUBTOTAL		88	103	122	166	166	169	166	168	168
Weight Class 4										
Truck Tower	031	5	5	3	2	2	4	4	4	4
Aerial Bucket Hvy.	040	7	6	6	7	6	3	3	3	3
Aerial Bucket Hvy.	045	160	170	177	189	196	187	187	190	189
SUBTOTAL		172	181	186	198	204	194	194	197	196
Weight Class 5										
Truck Heavy	027	48	48	52	63	64	67	70	70	70
Truct Tractor	029	9	8	9	9	9	9	9	9	9
Rotary Der. Heavy	055	67	61	60	42	32	27	28	27	27
Dig/Der. Heavy	065	40	56	53	88	88	86	86	86	86
SUBTOTAL		164	173	174	202	193	189	193	192	192
Powered Vehicles TOTAL		1259	1258	1343	1433	1440	1439	1400	1392	1387
Trailers & Misc.	076	509	510	520	537	533	553	512	504	498
SUBTOTAL		509	510	520	537	533	553	512	504	498
Construction										
Forklift	082	49	49	49	49	49	48	47	46	46
Backhoe	084	29	29	29	29	28	28	28	28	28
Trencher	088	20	20	20	19	19	19	19	19	18
Bulldozer	090	10	8	8	8	8	8	8	8	8
Track Vehicle	097	3	3	3	3	3	3	3	3	3
SUBTOTAL		111	110	109	108	107	106	105	104	103
TOTAL		1879	1878	1972	2078	2089	2098	2017	2000	1988

Figure 1-3
Fleet Inventory

Class 1: GVW 6,000 lb and less. These are compact, light-duty, full-size pickups, and light-duty passenger vans (5–8 people). About 80% are used for some personal service. Horsepower range: 100–160 hp (gasoline).

Class 2: GVW 6,001-10,000 lb includes heavy-duty pickups, forward-control chassis for walk-in vans, chassis cabs for plumbers, tire-service trucks with compressors, passenger-carrying vans (12–15 people), off-road four-wheel drives (called "utilities" by the manufacturers), and station wagons on truck chassis. About 20% see personal use. Range: 130–200 hp (gasoline), 130 hp (diesel).

Class 3: GVW 10,001–14,000 lb. Heavier step-vans and forward-control chassis for van bodies, ambulances, small fire trucks, and motor-home chassis. Range: 130–200 hp (gasoline), 130 hp (diesel).

Class 4: GVW 14,001–16,000 lb. Walk-in chassis—utility type work trucks, street light service, and mechanics' mobile repair vans.

Class 5: GVW 16,001–19,500 lb. Light-weight, medium-duty type chassis-cabs, farm and ranch flatbeds with stake racks, and conventional truck chassis for grocery delivery, etc. Range: 100–160 hp (gasoline), 130–190 hp (diesel).

Class 6: GVW 19,501–26,000 lb. Single-axle straight trucks, city delivery van-body types, one-way rental, school buses and farm trucks, and a few city delivery tractors. Range: 120–190 hp (gasoline), 130–225 hp (diesel).

Class 7: GVW 26,001–33,000 lb. Single-axle tractors and truck chassis for beverage bodies, tanks, city delivery vans and opens, and grain haulers. Range: 140–220 hp (gasoline), 160–225 hp (diesel).

Class 8: GVW 33,001 and over. Over-the-road linehaul tractors, heavy-duty trucks and tractors with tandem axles, dump and refuse trucks, and virtually all vehicles with linehaul or "big bore" diesels. Range: 200 hp and up.

Figure 1-4
Manufacturer Vehicle Classifications

agencies to gather comparative data on utility fleet performance (not only a utility compared to itself, but compared to other utilities). It also is a source for cost migration information as utilities change in size so that costs can be predicted and needs can be met.

The meaning of fleet costs must be understood. The numbers often are misleading until they are unbundled to identify the elements of cost. An oversimplified example is cost per mile. What makes up this number? Capital cost (i.e., principal and interest plus fuel), parts, shop labor, and registration costs? Or is it only parts and hourly payroll labor costs?

The difference between shop and payroll labor alone—$18 per hour for payroll and shop labor with indirect and overhead expenses at $60 per hour—will distort meaning, interpretation, observation, and, of course, any semblance of comparative data.

Add to this the clearing account concept, in which all transportation costs are spread to internal customers by an accounting system code definition. Here is an example of this concept. If the total cumulative cost for all chassis is $100,000 and there are 100 repairs completed, that is a cleared cost of $1,000 per repair rather than 1,000 repairs at $100 per repair. All the dollars are changed based on the number of repairs to a zero level.

In the clearing account concept, it is difficult to have observers with different perspectives conclude performance standards from this information; however, whoever is looking, measuring, watching, and paying attention to cost data can obtain some divergent observations. This often confuses the listeners who have different perspectives based on their observations and experience, which leads to multiple conversations comparing apples to oranges to peaches.

Fleet managers should be proficient in each category of cost topics in cost discussions and listen carefully to people's perspectives, interpretations, and conclusions. They should stick to the cost elements of the person's view, defining their meanings and clarifying terms within the category explained, based on all elements considered.

Fleet managers are best qualified to do this because fleet managers provide the input of cost information in each of the cost centers through purchase orders and other validated data input gathered prior to cost closeouts. Then, immediately after closeouts, the fleet managers should prepare graphic displays of cost summaries and give this information to customers, both internal and external, before they receive the information from the cost center and are left to develop conclusions through observations rather than facts.

The fleet managers should sort it out, define meanings and interpretations, and compare quarter to quarter (using the same number of days), last year's figures, month to month, year to year, with a published conclusion to each customer.

Proactive analysis with fleet managers taking the lead allows others see the observations and reasons why in measuring, watching, and paying attention to meaningful indices. Fleet managers define what is meaningful.

If customers are allowed to observe and selectively define their observations, fleet managers become reactive rather than proactive. The fleet managers must be on the offense—lead, rather than be led—lest they give control to their customers.

Fleet managers should demonstrate control of costs, show the push and pulls, and describe the process. If customers are allowed to take the lead, they will measure what is meaningful to them, many times using convoluted logic to ensure that the numbers say what they want them to say.

Show them, educate them, and organize information comparing apples to apples. Do this for them; lead them. I would rather prepare a menu for our customers to choose the service from us, and then serve them appropriate information. Fleet managers determine the information that customers need—not want—and educate them on what is important. Lead them. That is what managers do per Webster's definition. It is " . . . to do . . ."

Budget Timing

The budgetary process usually starts one fiscal year ahead of the present year, with the measuring at present vehicle inventory and utilization and projecting future needs. When the numbers and types of vehicles are agreed on for replacement, addition, or reduction, a cost is defined. That cost in annual and monthly dollars is put into consideration for the coming fiscal year budget for approval, that is, capital budget for straight purchases and operating or expense budget for leasing.

Vehicle Utilization

This process includes vehicle utilization in terms of hours, days, mileage, and fuel use to verify the vehicle operating history and comparison to company needs. Inner-city statistics will show low mileage, high fuel, hours, and day use. Suburban use will show higher mileages per fuel, hours, and day use; rural or highway use will show yet higher mileages for fuel, hours, and day use. It is extremely important to examine user environments and their impact on vehicle use and costs. Inner-city vehicles will show higher costs for mileages and tend to be replaced with lower miles. These miles are harder miles than highway or rural miles, which show higher costs at higher accumulated mileages. This is due to stop-and-go traffic patterns and congestion in the inner-city environment as compared to rural and highway applications.

Figure 1-5 shows the percentage of average class utilization based on time (days-hours) and should be discussed with each user to arrive at an acceptable level of vehicle and equipment utilization. The goal here is to replace a utilized vehicle. If the vehicle is not being properly utilized in the matter of efficiency, then bring that to the attention of the user to review the needs of the vehicle. If underutilized, then combine it with use of another vehicle to provide an economically acceptable level of use and resultant excessing of one of the two vehicles. To replace an underutilized vehicle proliferates a counterproductive environment.

An example of a snapshot of vehicle utilization is illustrated in Figure 1-6. A location is identified, each vehicle at that location is listed small to large, capacity is noted, and route number is assigned. The 24 hours are divided into 15-minute sections with use time noted in terms of hours and miles.

This process quantifies a 24-hour snapshot of the utilization of a fleet, vehicle by vehicle. It provides management with a format to review potential consolidations improving fleet levels, mix, and density requirements. In this format, this utilization information will provide targets or goals to be set for each type and class of vehicle. Obviously, specialty vehicles may have a low utilization level that is understood and acceptable, but they may not be candidates for replacement.

Figure 1-6 shows Routes 3 and 4 can be combined and one vehicle used, excessing one vehicle. With Routes 15 and 16, a similar situation exists, which results in the eight assigned vehicles being reduced to six vehicles, increasing fleet productivity.

The user and the transportation people can plan together to facilitate efficiencies in support of the goals and objectives of a company. This is the cornerstone of vehicle specification and replacement. The user is a customer,

DESCRIPTION	CLASS	WEIGHT CLASS 1 3 MO. AVG. 2001	UTILIZATION
Executive	01	12	78.4%
Standard	02	0	—
Intermediate	03	23	66.9%
Compact Sedan	04	182	78.3%
Station Wagon Int.	06	0	—
Compact Sta. Wagon	07	80	84.1%
Subcompact	08	0	—
Station Wagon	09	0	—
TOTAL		297	
		WEIGHT CLASS 2	
Van Compact	010	36	67.9%
Utility Veh. 4x4	012	20	93.1%
Utility 4x4 Com.	022	1	84.1%
P/U 4x2 Compact	013	287	89.1%
Com. Wagon 4x2	015	7	79.4%
Com. Wagon 4x4	016	12	94.6%
Van Small	017	125	79.4%
Pick-Up 4x2	018	62	81.1%
TOTAL		550	
		WEIGHT CLASS 3	
Pick-Up 4x4	019	41	69.3%
Van Walk-In	021	22	67.3%
Truck Light	023	25	52.2%
Truck Medium	025	10	72.4%
Aerial Bucket Sm.	032	68	87.3%
TOTAL		166	
		WEIGHT CLASS 4	
Truck Ladder	030	1	—
Truck Tower	031	4	74.2%
Aerial Bucket Med.	040	3	55.0%
Aerial Bucket Hvy.	045	187	64.4%
Rotary Der. Med.	050	0	—
TOTAL		195	
		WEIGHT CLASS 5	
Truck Heavy	027	70	44.2%
Truck Tractor	029	9	60.5%
Rotary Der. Heavy	055	28	62.6%
Dig/Der. Medium	060	0	—
Dig/Der. Heavy	065	86	52.1%
TOTAL		193	
Powered Vehicles		1401	
Trailers & Misc.		617	
TOTAL		2018	

Figure 1-5
Average Utilization by Vehicle Class

VEHICLE USE PLAN

FOR REGIONAL USE ONLY — APPROVED (Date and) — DATE OF — LOCATION — PREPARED BY

VEHICLE NUMBER	SIZE	ROUTE NO.	...	HOURS OPER	MILES OPER	REMARKS
123	1/4	3		6	28	Daily
124	1/4	4		6	40	Daily
222	1/2	8		7	50	Mon-Wed-Thurs
223	1/2	9		7	25	Daily
315	3/4	15		7	60	Daily
316	3/4	16		7	50	Daily
622	5	35		14	120	Daily
623	5	36		11	140	Daily
TOTALS				65	513	

Figure 1-6
(Courtesy U.S. Postal Service)
Snapshot—24-Hr Utilization
Form #4569

and a customer's real needs must be met. Transportation must review vehicle historical utilization data and share it with the user. This is the check and balance necessary for efficiency.

Vehicle Specifications

To purchase a vehicle, a vehicle specification is written. It is a component-by-component description of the vehicle that is to be used as a tool to perform a specific task. There are two types of specifications: functional and technical.

Functional Vehicle Specification

A functional specification is a specification that outlines the performance requirement of a vehicle. For example, the vehicle will carry 14,000 lb of stone, it will go 50 mph loaded, and it will climb a 12-degree hill loaded with 14,000 lb of stone at 40 mph. A functional specification puts the responsibility and liability on the manufacturer of the vehicle to assure that the user will have the vehicle that is needed to do the functional job required.

Technical Vehicle Specification

A technical specification is a specification that defines the entire assembly of a vehicle with component specifications, design of operating systems, and detailed systems performance to accomplish the user's need. The primary responsibility and liability reside with the author of the technical specifications, not with the manufacturer of the vehicle. Of course, the manufacturer would not provide a vehicle, system, and component that were not efficient. The question is, will the vehicle perform efficiently in its intended environment? The benefit of a technical specification is that it would provide users with a vehicle that would meet their needs more efficiently because it is tailored to their specific environment by the users.

Initiation of the Bid Process

When a specification is written and approved by the user, the transportation department, and the manufacturer, the bid is published for the manufacturers to submit costs to supply this vehicle. A meeting with the bidders is desirable prior to sending out the specification for a price quote. This allows for clarification to the bidders by discussion of the vehicle and its components, plus the expectations of the purchaser in terms of performance, life cost to operate and maintain the vehicle, and warranties, both expressed and implied.

The written specification, because of its detail, and a manufacturer's ability to exactly meet the required detail requires some clarification and explanation, and discussions are required prior to bid price responses. The companies bidding for the vehicle assembly can better provide for the user's need by better understanding the user's requirements.

Solicitation of the Bids

With the specifications written, the bid process requires the outlining of the customer's specific expectations. Several criteria must be defined, including delivery times, assembly sequence, chassis body and mounted equipment assembly timetables, warranties, penalties and latent defect qualifications, and, of course, payment terms. Other standard purchasing criteria and terms must be included, with the date, time, and format for quotes to be returned.

The goal is to obtain many qualified firms to bid stimulating competition, which results in the best overall price. If one brand is required, gather as many dealers of that brand to compete for the business. This ensures a competitive environment and posturing to allow negotiation in areas that need negotiation.

Having each of the components as a separate line item makes it possible to decide if costs are in line from each bidder. Items can be deleted to bring costs in line if necessary. Other considerations include multi-year pricing, reduced pricing for quantities over a certain amount, and penalty clauses for failure to deliver on time, combined with a bonus for early delivery.

Bid Evaluation

When the bids for a functional specification are received, the best value must be determined. Lowest price is important only for a technical specification. For a functional specification, best value should be the goal.

The responses must be evaluated to determine if they are technically correct, and that the respondents are reliable and can support their products in the future while in service. The bidders should be evaluated for their facilities, materials, processes, personnel, the client's historical experience with their products, and their prices. These should be reviewed as to successes and failures. Also, other users of this equipment should be consulted to explore their experiences with this manufacturer and their products.

Bid Award

With the technical, commercial, quality, and price evaluations in place for a functional specification, the bids can be prioritized for award of the project to the best value vendor. The best qualified bidder would be invited to a pre-award bid meeting to discuss each line item and its cost to ensure the bidder meant what the bid said. This meeting would provide an opportunity to qualify any discrepancies and questionable areas that are not clear, such as delivery times, penalties, and the manufacturers' defined warranties and their support services.

When satisfied with the vendor's proposal, the bid would be awarded to the best-qualified vendor. Should the first vendor not be clear in its technical, commercial, quality, and price, a meeting with the second or third vendor may be required until the requirements of the specification are met with its bid quote. In the case of a technical specification, the lowest price is the sole determination of an award.

Coordinating Manufacturing and Assembly Processes

When the bid is awarded and timetables are established, the targets and milestones would be noted and inspections initiated to measure progress. It is important to monitor progress because the replacement process is predicated on a timely delivery. This allows for the removal of the old vehicle before excessive maintenance costs are incurred, which would upset economic efficiencies. Not only is the user expecting the vehicle on a delivery date agreed upon, but the maintenance department could be deferring unnecessary repairs that would impact wasteful spending.

Any delays caused by material availability, labor deficiencies, and unexpected events must be resolved to ensure a timely delivery of a functional vehicle.

Pre-Paint Inspections

When the vehicle is assembled and tested to work properly, the end user, the customer, should be brought in at the manufacturer's expense and should inspect the unit at the manufacturer's facility to ensure that the manufacturer's workmanship is acceptable. By inspecting an unpainted vehicle, areas in question can be identified from a written summary of observations at the pre-paint inspection with copies to the delivery point garage for reference. Problems can be resolved at that time, reducing delivery and in-service delays. This is an especially desirable process should you be purchasing more than one of this type of vehicle. Problems can be corrected on the first vehicle, ensuring the following units have the unit integrity needed. Should a defect be identified when the vehicle is in service a period of time, repairing the remainder of the order after those vehicles are painted is more costly to both parties. The manufacturer incurs travel and external facilities expenses, the fleet loses availability, and the user has doubts about more problems that could be expected.

Final Vehicle Inspection

After the pre-paint problems are corrected and the first unit is painted, the delivered unit can be inspected by the end user before shipment or after shipment, depending on the severity of the pre-paint problems identified, at the expense of the manufacturer.

The goal is to accept the vehicle when delivered at the same time the old vehicle is scheduled to be replaced. This allows the old vehicle to be turned in upon receipt of the new vehicle after it is prepared for service by the in-house garage or the chassis dealer. This timely and efficient turnaround allows for a final budgeted payment, sale of the old vehicle, and the fleet size being controlled.

Life-Cycle Costing

Replacing a vehicle is an annual economic decision. The key factors to compare in evaluating an efficient replacement program are:

- Company growth, reduction of work, *status-quo*
- Old and new vehicle principal and interest (depreciation)
- Old and new vehicle maintenance cost
- Old vehicle resale value and market conditions
- Old and new vehicle operating cost
- New vehicle manufacturer incentives—buy backs—credits

When the principal and interest of the old vehicle decrease, the maintenance and operating costs usually increase. You should measure the increase in maintenance and operating costs to see if it is more or less than the principal and interest decrease. The annual resale value information proves to be the tie breaker, with fleet incentives made available by the manufacturer each year.

If the resale value of the old vehicle is high enough, coupled with the incentive for the new vehicle, it will reduce the principal and interest of the new vehicle. If the principal, interest, maintenance, and operating costs of the old vehicle become higher than the principal, interest, maintenance, and operating costs of the new vehicle, the old vehicle should be replaced.

If resale values are low, with low manufacturer incentives, then this would support keeping the old vehicle and taking advantage of minimal principal and interest costs after the vehicle is depreciated. In fact, there will be a higher maintenance and operating cost due to the extension of the replacement cycle of the vehicle. Determine if the increased maintenance and operating costs exceed the reduced principal and interest cost savings. In the 1990s, the manufacturers started to provide advanced and improved power and vehicle designs to support extended vehicle life cycles. The defects were corrected by the mid-1990s, and the increased vehicle performance helped end users benefit from these efficiencies.

This is an annual review for all vehicles because of variable economic climates impacting company workloads, vehicle resale values, manufacturing incentives, available capital, and interest costs.

Most older, larger vehicles do not have high resale values or manufacturer new vehicle incentives to impact life-cycle economics. If they do because of economic and special applications, this should be included in a life-cycle annual analysis.

Older vehicles should be kept as long as their total cost is less than the total cost of the new vehicle. This includes mounted equipment for vocational trucks to support improved work methods.

Vehicle Disposition

Selling of a vehicle carries a liability with it. This vehicle must be usable to obtain a good sale price. Should it be defective, a decision must be made whether to salvage it rather than sell it. It should pass the state inspection and have a stated warranty published. If a warranty is not published, the company is liable for an implied warranty that probably would be greater in content than can be supported by the company.

A detailed inspection of the vehicle should be made, pictures taken of unique items, and the condition noted. Disposal can be through a retail, auction, trade-in, or wholesale process, depending on the market position at the time of sale.

Replacements should be timed to recognize new model introductions. Because old models get older as new models are introduced, the resale value of the old vehicle is lowered.

Warranties

The company should define its warranty expectations when it solicits for bid responses. Whether it accepts the manufacturer's warranties or defines its own, this should be clear prior to an award to the successful bidder. Upon delivery, the terms and conditions should be known and the company prepared to follow up and enforce them. Manufacturers make mistakes, and the customer should be made whole for the deficiencies.

An initial target of 5% of the purchase price of each vehicle should be established for warranty return during the first year of service. This target should be based on past experiences. It could be exceeded or fall short, depending on each model year. It is important to set a target and measure performance. We will find more opportunities for reclamation from the mounted equipment of a vocational truck than from the chassis.

Final Payment

When the vehicle is delivered, accepted, and put in service, then payment should be made. The terms for payment should be delineated in the solicitation process and reviewed prior to award.

Note that expectations are for a vehicle to be delivered in a turnkey condition. A vehicle delivered in an in-service condition should be paid for. To pay for a vehicle because it is delivered is not desirable. The old or replaced vehicle must be turned in when the new vehicle is put into service. If the new vehicle is not reliable and not turnkey, it is more difficult to get the old vehicle turned in within a timely manner.

Partial payment will be necessary to obtain ownership papers to license and register the vehicle so that it can be put into service and shaken down in a reasonable time period. Thirty to sixty days of use is a reasonable time frame.

Final payment is contingent on a vehicle that is completely operable and able to be put in service for the user to turn the key and use the vehicle efficiently as a tool. You should hold the manufacturer accountable through warranty latent defect and liquidated damages, with terms and conditions agreed upon in the delivery process.

These areas will be expanded in detail in the following chapters, using cost-effective, practical, logical analysis to support an efficient vehicle replacement process.

Chapter 2

Vehicle Utilization and Cost Strategies

⊹

Vehicle Utilization

If a vehicle is not being properly used, it should be removed from the fleet. The best time to do this is when the vehicle is being considered for replacement. Why replace an underutilized vehicle? At this time, it is possible to identify and communicate what is the use of the in-service vehicle in terms of hours, days, miles, and fuel use. The desired utilization for this vehicle in its assigned environment must be isolated. This action is best received when considering replacement. People will listen and be receptive to cost analysis at this time.

The vehicle replacement program is defined generally in age (years/months) and mileage. This point of reference is determined based on cost of new versus cost of old. When the new vehicle costs less to own and operate than the old vehicle, then it is time to replace the old vehicle with the new vehicle. If the old vehicle costs less to own and operate than the new vehicle, the alternative is to keep the old vehicle in service. On an annual basis, this cost analysis is performed, and the company replacement program is adjusted to reflect any changes.

Some questions to consider are:

- What is an acceptable utilization rate for each class/type of vehicle in a specific location?

- How does each in-service vehicle compare to this rate?

- How many spare vehicles will be kept in the fleet to cover annual accidents that result in a totally unrepairable vehicle?

- What are effective utilization targets for a personally assigned vehicle, pool vehicles, department-assigned vehicles, job-assigned vehicles, maintenance replacement vehicles, and specialized vehicles?

With company-accepted targets for each location, you can proceed to set a policy on desirable utilization rates and on an annual basis calculate present usage in terms of targeted usage. Vehicles not meeting usage targets should be reviewed and not replaced. These vehicles should be removed from service. Those vehicles meeting utilization targets will be replaced when they meet their replacement criteria.

To track utilization during the life of vehicles, a monthly cost report of each vehicle should be kept to advise the users what their individual costs are. These costs should be related directly to use. These costs should be grouped into fixed and variable categories.

Fixed costs should include principal and interest (depreciation—wear and tear due to miles and time), registration and taxable costs, permits, and other related fixed annual costs.

Variable costs include fuel, labor, parts and supplies, accident costs, tolls, parking, and traffic violation fees.

These costs should be summarized monthly and shared with the users in terms of utilization criteria.

For example:

Vehicle 1—Compact Car

Fixed costs	$200 per month
Variable costs	$300 per month
Total costs	$500 per month
Mileage	2,000 miles per month
Cost per mile	= 25¢

Vehicle 2—Compact Car

Fixed costs	$200 per month
Variable costs	$200 per month
Total costs	$400 per month
Mileage	1,300 miles per month
Cost per mile	= 31¢

If the environments of both vehicles are the same and the acceptable range of costs is 24¢ to 27¢ per mile, Vehicle 1 is acceptable in its cost. Vehicle 2 is questionable, and investigation should be initiated to determine the cause. Is it temporary due to operating conditions? Do its cumulative costs average in the range of 24¢ to 27¢ per mile?

Vehicle 3—Compact Car

Fixed costs	$200 per month
Variable costs	$100 per month
Total costs	$300 per month
Mileage	500 miles per month
Cost per mile	= 60¢

Vehicle 3 is twice the average cost target. If the cumulative cost for the last 6 to 12 months supports the 60¢ per mile, then this vehicle is a candidate for removal from the fleet and/or reassignment to a higher use area. If this vehicle is scheduled for replacement and no higher use area exists, then it is a candidate for removal from the fleet without replacement. If a major component was just replaced (e.g., engine, transmission), the vehicle is OK for another two to three years. This decision should be made prior to the major repair being authorized, using 30% of its residual value being targeted as a maximum maintenance expense.

Each class of vehicle, car, light truck, bus, medium truck, heavy truck, trailer, or off-road equipment would have an appropriate utilization level established by historical information and operation needs. A goal here is to charge the using department with their usage costs and encourage the users to maximize their profitability through their efficient use of their vehicle resource.

Obsolescence

On an annual basis, replacements can be made for other reasons. Some of those reasons are:

A present vehicle no longer efficiently meets user needs.

A dump truck with a 5-yd capacity needs to move 100 yd of product per eight hours. A 5-yd dump truck would make 20 trips per 8-hr time period to move

100 yd of material. A 20-yd dump truck would make five trips per 8-hr period more efficiently. Replace the 5-yd dump truck with a 20-yd dump truck. No matter what condition the 5-yd truck is—it could be new—remove it, reassign it, or replace a worse truck and excess the worse truck.

A bus is picking up 20 passengers per run on an established route. The route generates 40 passengers per run. A larger 40-passenger bus is needed rather than using two 20-passenger vehicles that would require two drivers.

Improved Operating Efficiency

A full-size pickup truck gets 10 mpg. A compact pickup gets 20 mpg. The reduced fuel cost would support using the more efficient compact pickup truck as long as the carrying capacity of the compact pickup is acceptable.

A diesel-powered 40,000 GVW chassis with lower maintenance cost and increased fuel economy can provide more efficiencies than a comparable gasoline chassis, so opt for the diesel chassis.

Replacement Forecasting

On an annual basis, the present fleet should be projected for next year's usage. Delivery timetables would be forecasted and budgeted for this replacement cycle. An example would be as follows:

The fiscal year runs from January to December. Based on historical usage in miles, in March of the previous year, the company would forecast usage into the coming year. Picking a delivery date based on an efficient new model order cycle, September 1, of the future year would be targeted as a replacement delivery date. Forecasting the accumulated mileage of the vehicle as of September 1 would identify in descending order those units that qualify for a mileage replacement. Presume 100,000 mi is the qualified replacement target. All vehicles that meet 100,000 mi in September would be highlighted in descending order. A determination of those vehicles under 100,000 mi that would meet the 100,000-mi target in the next fiscal year would be highlighted. A decision would be made as to how many of those vehicles would qualify for replacement, and they would be added to the list. A cost per vehicle would be assigned to them, and this total cost would be budgeted for the coming fiscal year. The strategy is to level the annual purchasing for company planning programs.

An assessment of the total need of the company would be made based on business plans, and a verified amount would be agreed upon. This budget would be a basis for ordering vehicles for replacement.

Replacement Alternative

There are other alternatives to a vehicle replacement program:

- Rebuilding a vehicle
- Reconditioning a vehicle
- Remanufacturing a vehicle
- Relifing a vehicle

Definitions of Rebuilding and Relifing/Updating

Are relifing/updating costs: operating and maintenance costs?
                                 ...or...capital expenditure?

Probably capital expenditure, depending on the accountant's point of view.

Capital cost for depreciation = book value at time of relifing/updating + cost of relifing.

Therefore, in practice, the overhauling of equipment cannot be divided into conveniently spaced and defined steps but is a gradation, rising in cost and complexity as the rebuild becomes more thorough. The essential difference between the various degrees of rebuilding and what has been called "relifing" is the incorporation of new technology to improve the productivity of the machine and the operator's working environment to today's levels or standards.

A rule of thumb says: do rebuild if you get three-quarters the life usage of new with one-half the cost of new as your rebuilding target.

For example:

Vehicle Cost: $70,000 new
This vehicle at 10 years old needs $9,400 of repair. Its depreciated value is $10,000.
Life cycle shows major repairs needed now that are more than 30% of the residual value of the vehicle.
Should we rebuild, repair, remanufacture, or replace?
Goal: If we were to successfully rebuild, we would spend one-half the cost of new; we would look for three-quarters the life of the new vehicle to replace a $70,000 vehicle that costs $100,000 new today.
One-half of $100,000 = $50,000
New $70,000 vehicle lasted 100,000 mi = 70¢/mi
Rebuilt @ $50,000 vehicle lasts 75,000 mi = 66¢/mi

In this case, $50,000 to build or relife this vehicle is a justifiable expense as long as the configuration of the vehicle supports the present work methods.

In this industry with the type of equipment used, the term "relifing" should be used only when a manufacturer takes an old (5–10 year) machine back to the factory, renews it, and returns it to the owner with a "new machine" warranty.

Rebuilding: Synonymous with "overhauling"— work performed inside or outsourced, which involves taking a complete machine or a major component part of a machine and disassembling it completely; cleaning; repairing and, if required, replacing any failed parts; testing; and reassembling.

Reconditioning: Work performed inside or outsourced, which involves taking a complete machine or a major component part of a machine and disassembling it completely; cleaning; replacing all excessively worn parts, regardless of their condition; testing; and reassembling.

Remanufacturing: The same as for "reconditioning," except that the machine or major component is shipped to the manufacturer or his agent, where either the work is done on your component or an exchange is made for your component. In either case, the component carries the manufacturer's guarantee of a level of new quality (usually stated in time).

Relifing: Work in your own facility or in the facilities of an outside company (manufacturer, agent, specialist), which involves taking a complete machine and disassembling it completely; cleaning; replacing all parts (including obsolete parts with current upgraded equivalents); and incorporating the following into the reassembling:

1. All the manufacturer's product improvements

2. Viable design changes resulting from operating experience and/or technological change and/or the requirements to enhance performance

3. Current legal or company-required health, safety, acoustical, and vibration standards

4. Parts and/or sub-component assemblies that can be viably substituted on the machine and that will result in a reduction in the number of items held in parts inventory

If the work is done by an outside company, then both engineered changes and overall performance will be guaranteed at the same level as the original machine. The completed machine will have the same life expectancy as a new machine.[1] The cost should be no greater than the new cost—preferably less than new cost.

Used Vehicle Alternatives

Vehicle rental agencies for light, medium, and heavy vehicles have excellent purchasing power. Their replacement cycles are based on low maintenance costs. Their goal is to buy a vehicle and use it until it needs some basic maintenance. At that point, they will sell it rather than maintain it.

As a fleet operator, you must look at these vehicles as a source to replace fleet vehicles. Because in-house maintenance facilities or vendors are available to

---

1. Caution: Original equipment manufacturers may not recognize the work of your company or the work of an outside company when refurbishing their vehicle and equipment and updating it to current standards because they cannot be assured the work is done to their standards and that their approved parts were used. Therefore, liability is an issue.

repair the vehicles, it is possible to efficiently operate and maintain these vehicles. The combination of the rental agency's lower purchase price of these vehicles when new and the 30% depreciation registered after the first year in use provides an attractive negotiated purchase package to be considered in replacing older vehicles.

For example:

You would spend $20,000 to purchase a compact car on a fleet purchase program.

A rental company would purchase the same new compact car one year ago for approximately $12,500 because of a large volume discount. Looking at a 30% depreciation at the end of one year, this $12,500 vehicle would be worth $8,750 at one year old and 20,000 mi on it.

You could buy a new vehicle at $20,000 or a one-year-old 20,000-mi vehicle at somewhere between $10,000 and $14,000. This is a win-win for both parties. It is worth the time invested to evaluate this alternative.

Interest

If money is loaned or borrowed, a fee is charged for its use. That fee is called interest. Anyone who has borrowed money to purchase a house or car is familiar with the process.

There exists a very explicit set of mathematical rules that govern the relationship between the amount borrowed (or loaned, depending on your point of view), the interest rate, the duration of the loan, the interest owed, and the method of repayment. For example, if you borrow $1,000 at 12% interest and agree to repay the $1,000 plus interest one year from now, the amount due will be $1,120 at the end of the year. If you prepay the interest, you will receive $892.90 now and pay back $1,000 at the end of the year. You can also repay the loan by making 12 monthly payments of $88.80, totaling $1,065.60. In the language of economic analysis, $1,000 now is equivalent to $1,120 a year from now at 12% interest, or $1,000 now is equivalent to 12 monthly payments of $88.80. This is a simple illustration of the fact that time affects the value of money.

The simplest of the equivalence relationships is the "compound amount" formula that is used to compute the relationship between a single present amount and an amount in the future. For example, we have said that $1,000 now is equivalent to $1,120 one year in the future at 12% (the interest rate). In two years, the future amount would be $1,120 + (0.12 × 1,120) = $1,254.

1 year = $1,000 × 12% = $1,120

2 years = $1,120 × 12% = $1,254

However, to deal with a more complex problem, such as finding the equivalent value of the after-tax cash flow, more powerful tools are required. In engineering economic analysis, cash flow diagrams are used to organize the problem for analysis. Interest rates for various instruments (e.g., commercial paper, Treasury bills, bonds, and other sources) are published weekly by the Federal Reserve and can be secured by subscription. (Call 202-452-3206 for information and follow the telephone prompts for the desired information; see Figure 2-1.)

These data are released each Monday. The availability of the release will be announced when the information is available on (202) 452-3206.

H.15 (519)For immediate release October 2, 2001

SELECTED INTEREST RATES
Yields in percent per annum

Instruments	2001 Sep 25	2001 Sep 26	2001 Sep 27	2001 Sep 28	2001 Sep 29	This week	Last week	2001 Sep
FEDERAL FUNDS (EFFECTIVE)[1]	9.09	9.05	9.03	9.20	9.24	9.02	9.05	9.02
COMMERCIAL PAPER[2][3]								
1-Month	8.88	8.90	8.92	8.92	8.99	8.92	8.84	8.87
3-Month	8.71	8.74	8.77	8.78	8.83	8.77	8.65	8.70

[1] Weekly figures are averages of seven calendar days ending on Wednesday of the current week; monthly figures include each calendar day in the month.
[2] Quoted on bank-discount basis.
[3] Rates on commercial paper placed for firms whose bond rating is AA or the equivalent.

Figure 2-1
Federal Reserve Statistical Release

33

Inflation

Inflation is the loss in value of money over time due to the higher costs of materials and labor in making goods (Figure 2-2).

1965	1.9%	1978	9.0%	1991	5.0%
1966	3.4%	1979	13.3%	1992	5.0%
1967	3.0%	1980	12.4%	1993	5.0%
1968	4.7%	1981	8.9%	1994	5.0%
1969	6.1%	1982	3.9%	1995	5.0%
1970	5.5%	1983	3.8%	1996	3.0%
1971	3.4%	1984	4.0%	1997	3.5%
1972	3.4%	1985	3.8%	1998	4.0%
1973	8.8%	1986	1.1%	1999	3.0%
1974	12.2%	1987	4.4%	2000	3.5%
1975	7.0%	1988	4.9%	2001	3.2%
1976	4.8%	1989	4.9%	2002	2.5%
1977	6.8%	1990	5.0%	2003	2.5% est.

Figure 2-2
Inflation Summary

Depreciation

Depreciation is the loss in value of the vehicle during the time it is owned due to:

1. Passage of time
2. Wear and tear
3. Miles the vehicle is driven

Equipment and vehicles - 5 years
Depreciation is the salvage value (sale price) minus the original cost

4-Year Sample:

Example - Cost new	$20,000	
4-year-old salvage value	7,000	(Estimated market value)
	$13,000	= Depreciation loss in value

Calculated salvage value = Loss of 30% first year + loss of 20% of the residual value each of the following years

4-year-old vehicle = 70% × 80% × 80% × 80% = 36%

$$\$20,000 \times 36\% = \$7,200 \text{ salvage value}$$

$$\text{Depreciated value} = \$12,800$$

Salvage value = $20,000 vehicle = 4-year-old vehicle = 36%

70% × 80% × 80% × 80% = 36%

30% loss first year, 20% thereafter each year
$20,000 × 36% = $7,200   4 years
$20,000 × 18% = $3,600   7 years
$20,000 × 9% = $1,879   10 years

Types of Depreciation

4-Year Sample:

Straight Line

Straight line depreciates 4 years

$$\$20,000 - \$7,200 = \$12,800 \div 4 \text{ years} = \$3,200/\text{year}$$

Accelerated

Double declining balance (accelerated) depreciates 4 years

$20,000 – $7,200 = $12,800 is maximum allowed for 4 years

$20,000 ÷ 50% = $10,000 first year

$12,800 – $10,000 = $2,800 second year

35

Sum of the Years' Depreciation—4 Years

$1 + 2 + 3 + 4 = 10$

$$\$12,800 \div 10 = 1,280 \times 4 = \quad \$5,120$$
$$1,280 \times 3 = \quad 3,840$$
$$1,280 \times 2 = \quad 2,560$$
$$1,280 \times 1 = \quad \underline{1,280}$$
$$\$12,800$$

Capital Expenditures

When you have to find money for new capital expenditures (for fleets, that's mostly for new trucks), there are only two places to go:

1. Retained past earnings (after-tax profits) and depreciation
2. The "money market"; take on debt or sell more stock

Say you bought a new pickup truck for $25,000 in 1997. Uncle Sam said you could exempt from taxes an amount of your income in each of the next five years totaling $25,000. (Figures and times are very approximate and are as an example only.) Now in 2002, you have set aside more than $25,000 tax-free dollars. But by 2002, the same truck, plus the required extra features, costs $45,000. That extra money must come out of retained earnings (which most companies might not have, especially after a poor profit year) or from outside sources.

To take on more debt at a time of high interest rates reduces profits in future years. To sell new stock is difficult during a time of lean profits and dilutes the interest of present owners. Therefore, in times of inflation, the government's limiting depreciation to "100% of original cost" can force companies between two unappealing capital-raising alternatives. In such times, leasing may prove to be a way out.

Obviously, not every fleet manager must become an expert on these intricate financial and tax matters. However, these are coming to the forefront, and it would behoove all to gain enough understanding to avoid being out in left field when money managers and top executives start talking about this.

The Cost of Borrowing Money

Annual interest on a loan is calculated by the formula

$$I = (A - 5.5B)R$$

where I = The interest charge per year
  A = The amount of the loan
  B = A ÷ the number of months of amortization
  R = Interest rate
  5.5 = Interest-principal curve leveling factor

Example:

If a truck costs $50,000 and was paid for with an 18% loan amortized over 100 months, the annual interest would be

$$I = \left\{ \$50,000 - \frac{(5.5)(\$50,000)}{100 \text{ months}} \ 0.18 \right\}$$

= $8,505 per year, or $708.75 per month

Monthly payments consist of the monthly principal, B, plus interest, and in this case would be

$$\frac{\$50,000}{100 \text{ months}} + \$708.75 \text{ interest} = \$1,208.75 \text{ total payment per month}$$

Thus, the total cost of the loan would be

$1,208.75 × 100 months = $120,875

Using similar calculations, if a loan for the same amount were amortized over only 36 months, the total cost—interest plus principal—at maturity would be $72,864. This represents a 40% savings over the 100-month loan.

It is advisable for would-be tractor and truck buyers to shop for the lowest possible interest rate, then determine the shortest comfortable payback period to minimize the "interest bite." Figure 2-3 shows the total interest charges on

Years of Amortization	Total Interest Paid	% of Total Interest Paid on Value of Loan
1	$ 4,875	9.75%
2	$11,813	23.63%
3	$19,438	38.88%
4	$27,407	54.81%
5	$35,582	71.16%
6	$43,895	87.79%
7	$52,306	104.61%
8	$60,790	121.58%

Figure 2-3
Interest on $50,000 Borrowed at 18%
over Varying Amortization Periods

a $50,000 loan at 18% amortized over various time periods. Also shown is the percent of the principal that the interest actually represents.

Tracking the Cost per Vehicle

The effort to gather per-vehicle costs leads to multiple bonuses. The resultant data can predict vehicle life cycles, optimal times to buy, costs of scheduled versus unscheduled maintenance, and component costs, and can help in establishing priorities. Information can be forthcoming on a printout once a month or on a daily basis. Regardless of the frequency, we should have the capacity to fix costs on a per-vehicle basis.

Proper management demands not only planned spending and monthly tracking, but also the flexibility to respond to overages and underages in actual costs. This flexibility allows a manager to initiate corrective action to keep within budgetary restrictions.

Classifications of Vehicle Expenses

When projecting costs, a fleet manager is faced with a vast number of areas that require budgeting. The following three general expense categories are broken into their component parts, as an example of the detail desired. You need to define expense cost categories based on your company formats for consistency and for quarterly and annual comparisons to determine efficient performance activity.

Operating Expenses (expressed in cost per mile):

1. Fuel (include additives, if allowed).

2. Oil. Include lubricating oil, additives (if allowed), and oil changes. (Do not include lubrication or filters.)

3. Tires. Include the purchase of replacement tires and snow tires plus the cost of studding; deduct the expense for returned tires. Include repairs and charges for rotation. (Do not include balancing or alignment charges.)

4. Maintenance and repair. Include all mechanical and electrical repair, excluding repair of accidental damage. Include chassis lubrication, transmission and hydraulic fluids, filters, wheel balancing and alignment, brake adjustments, and towing. Include all charges for regular maintenance and tune-ups.

Standing Expenses (expressed in dollar cost per vehicle per month):

1. Investment cost. Include the total capitalized cost of the vehicle; deduct charges to employees for optional equipment.

   a. For leased fleets, use the monthly rental payment.

   b. For company-owned fleets, include the interest on money used for purchases.

2. Depreciation. Include reserves established at predetermined rates, as well as adjustments after cars are disposed of. If insurance payments have been received to cover accident repairs but these repairs were not made, the depreciation due to the accident should not be considered.

3. Insurance. Include all premiums for insurance coverage.

4. Administrative expenses, if available, should be included in the standing expense category. These would include the salaries of fleet department personnel, the share of corporate overhead (rent, light, heat, telephones,

pro rata costs of the accounting, legal, and finance departments, and other services needed for fleet operation).

Incidental Expenses (expressed in dollar cost per vehicle per month):

1. Licenses and taxes. Include city and state licenses; federal, state, and city taxes; personal property, sales, use, and rental taxes; registration and license fees; notary public fees; title and transfer fees; inspection fees; permits; and fees for duplicate titles.

2. Accident repairs. Include all labor, materials, towing, and other expenses resulting from an accident. Deduct from the cost of repairs payments from insurance companies or others, regardless of the date of the accident, including payments received for work not performed.

3. Washing. Include all company-paid expense for washing, waxing, and polishing.

4. Parking and tolls. Include all highway, ferry, tunnel, and bridge tolls. Include the use of outside parking lots, parking meters, garages, and garage allowances.

5. Miscellaneous. Include any items indirectly related to the operation of the vehicle not covered elsewhere, such as antifreeze, flares, seat covers, floor mats, and chains.

Glossary of Terms

Operating expense. All expense directly related to running a vehicle: fuel, oil, tires, repairs, and maintenance. These expenses reflect vehicle usage, are directly connected to the number of miles driven, and should be expressed as a cents-per-mile figure to provide a meaningful relation between mileage driven and dollar costs.

Maintenance and repair. All actions taken to keep company vehicles at peak efficiency at minimum operating costs.

Mechanic recruitment. From where do our mechanics come to repair vehicles and equipment?

Agricultural community—In the past, the agricultural community provided skilled mechanics

Military—Retired

Nepotism—Family

Associate–Friend

Shop out of business—Excess employees

Large shop to a small shop

Small shop to a large shop

Daytime hours—Weekday, opportunity to move from nights; weekend to weekday daytime hours

Travel time reduced or eliminated: be home regularly

Vocational education—Work study; cooperative education

Construction and transit industries, move to local shop

Better benefits

Better working conditions

Ethnic community more familiar to their ethnicity

Global conflict refugees escape persecution

Retired—Part-time mechanics; 2 people cover one 40-hr shift

Driver/equipment operator—Perform basic mechanic work

Standing expenses. Expenses incurred in vehicle operation that have little or no direct relationship to mileage driven, such as vehicle cost, depreciation, administration, and insurance. Items in this category are sometimes referred to as fixed or nonvariable expenses. They are predetermined and do not vary during the life of the vehicles. These expenses are best expressed in a dollar cost per vehicle per month.

Investment cost. The total amount paid for a vehicle. The cost of a company-owned vehicle should reflect the interest paid on funds used to purchase it. If the vehicle is leased, the cost should include the service charges by the leasing company. This item sometimes is known as capitalized cost.

Depreciation. The difference between the purchase price and the net disposal receipts for a company-owned vehicle.

Insurance. The protection purchased against claims of bodily injury or property damage resulting from accidents and against losses due to fire, theft, and vandalism. Premiums usually are based on previous experience and losses. Companies can "self-insure" against collision or medical losses by setting aside a reserve for each vehicle.

Administration expenses. The cost of internally servicing and managing the fleet of the company. In addition to direct salaries paid to fleet department personnel, this line should include corporate charges for specialized staff assistance and a *pro rata* share of corporate overhead (e.g., rent, light, heat, telephone).

Incidental expense. Certain expenses connected with vehicle operation do not occur on a regular basis, nor are they directly related to mileage or car use. These include licensing and taxes, accident repairs, parking and tolls, and miscellaneous. The recommended procedure is to present these expenses in dollars per vehicle per month.

Figure 2-4 illustrates various patterns of ownership cost, which must be tracked to guide management.

Figure 2-5 shows a sample budget and cost performance sheet, with direct and indirect costs under each management element listed. Budgeted expenses are

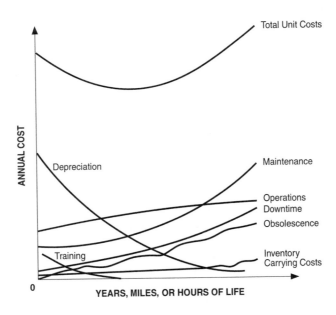

Figure 2-4
Patterns of Vehicle Ownership Cost
(Courtesy Byrd, Tallamy, McDonald, and Lewis, Falls Church, VA)

CONSOLIDATED EQUIPMENT OPERATING BUDGET AND COST PERFORMANCE REPORT

ORGANIZATIONAL UNIT _____

COST ITEM	MONTH OF			TOTAL TO DATE			SUMMARY TO SPEND		
	ACTUAL	BUDGET	VARIANCE	ACTUAL	BUDGET	VARIANCE	FORECAST	BUDGET	VARIANCE
Fuel Distribution									
Gasoline/Diesel Oil									
Oils and Lubricants									
Fuel Distribution Labor									
Fuel Transportation									
Commercial Fuel Purchases									
Distribution Facilities									
Maintenance									
Depreciation									
Shop Operations									
Direct Labor									
Direct Parts and Materials									
Commercial Repairs									
Transportation Charges									
Parts Distribution Costs									
Stockroom Labor									
Delivery Operations									
Parts Procurement Costs									
Miscellaneous Parts									
Shop Overhead									
Unassigned Labor									
Supervision									
Shop Supplies									
Utilities and Upkeep									
Shop Equipment									
Facility Depreciation									
Office of Equipment Management									
Salaries and Benefits									
Planning and Procurement									
Fiscal Management									
Parts and Materials									
Procurement									
Out-of-Pocket Expenses									
Support Services									
Insurance and Fees									
Capital Account Charges									
Depreciation of Equipment									
Amortization of Equipment									
Betterments									

Figure 2-5

Sample Budget and Cost Performance Sheet Outline

(Courtesy Byrd, Tallamy, McDonald, and Lewis, Falls Church, VA)

compared with actual expenses for each month, as well as for the year to date and the year end. Comparisons with the same time in the preceding year may also be made. Any variances between budgeted and actual expenses should be noted. On a monthly basis, variances should be addressed in a businesslike fashion, with corrective action initiated.

Because of changing economic situations in a company and the nation, you must be flexible enough to note counterproductive changes and respond positively to stabilize operations and prevent unnecessary losses.

As a fleet manager, you should be involved in the budget process because this is the area where you compete both offensively and defensively. Depending on the posture of your department in the company hierarchy, always be prepared to justify cost-saving technology. Discussion and planning must be factual. The dollar is the only term that makes all departments equal; the dollar is factual and unemotional. Past experience can be questioned, but accurate cost analysis is not subject to interpretation.

Leasing Alternative

A company planning to replace vehicles would have another need to use budgeted capital to expand or upgrade present property or other revenue-producing areas. Because you cannot use capital for two separate transactions, you need to obtain additional capital to do both.

Leasing the vehicle is a viable alternative to consider. First, here is an evaluation of a capital venture to expand operations.

Presume $5 million is needed to either purchase vehicles or build a factory. The factory will generate revenue to add to gross income. An oversimplified example is as follows:

$5 million builds a factory that generates $100 million in revenue. Presume net profit is $7 million (after paying all expenses, materials, labor, and overhead).

Presume the company pays a 50% tax on this net profit. Paying this tax on $7 million amounts to $3.5 million. This covers the manufacturing option.

Now the vehicle replacement strategy at the same time. $5 million is borrowed from a vehicle leasing company for three years, and a total of $3 million interest is paid. Total fleet cost for this replacement action is $8 million. The interest is paid $1 million per year for three years. The principal paid is $2 million in Year 1, $2 million in Year 2, and $1 million in Year 3.

Year 1	$2,000,000	Principal
	$1,000,000	Interest
	$3,000,000	Total Year 1
Year 2	$2,000,000	Principal
	$1,000,000	Interest
	$3,000,000	Total Year 2
Year 3	$1,000,000	Principal
	$1,000,000	Interest
	$2,000,000	Total Year 3

Total payment $8,000,000 3 years

Because the annual profit is $7 million for manufactured goods, you can deduct $3 million from the profit in Year 1. This means you will pay tax on $4 million instead of $7 million.

Tax on $7,000,000 profit is 50% = $3,500,000
Tax on $4,000,000 profit is 50% = $2,000,000

Year 1	Net difference is	$1,500,000
Year 2	Net difference is	$1,500,000
Year 3	Net difference is	$1,000,000
Total tax avoided		$4,000,000

In this oversimplified example, you have spent $4 million in three years for a $5 million fleet replacement. $5 million of the budgeted capital was used to expand the operation. The revenue generated from the expanded operation has offset all interest charges of the borrowed capital plus 20% of the principal cost of the fleet.

Equipment Replacement

Timely replacement of equipment is essential to an efficient fleet operation. It is important to base acquisitions on acceptable expense levels, and, as older vehicles exceed these levels in upkeep, decide whether to rebuild or replace. However, vehicle purchasing is a complex decision. Needs change, money costs change, and maintenance and repair costs change. As vehicles age, they require more maintenance and may become technologically obsolete. In addition to getting better fuel mileage, new trucks are more efficient machines, offering increased productivity and safer performance.

You should ask yourself certain questions when considering any truck purchase:

1. What is my current vehicle utilization rate?

2. Can I sell what is not being utilized and apply the proceeds to the purchase of new equipment?

3. When I purchase new trucks, will they be able to replace more than one truck, thus raising my fleet utilization rate? In other words, can I buy two vehicles and remove three old units? (The old units can be rebuilt, sold, or both, depending on business demands. If I shortchange myself, can I rent units to cover my peak demands or to buy time to purchase additional newer units, stabilizing my utilization rate?)

4. How many units are costing more than the average for their current class? What vehicles are about to exceed this cost? (This assumes, again, that the vehicles are being fully utilized.)

5. How many vehicles will my present shop maintain?

6. Which ones should I rebuild, and how long would it take?

7. What equipment should be rebuilt by a vendor, and how long would it take?

8. How long will it take to get delivery if I order new trucks today?

9. Can I lease with full service? Can I lease with partial service? Can I finance-lease and service the vehicles myself?

Leasing

Today a 100-truck fleet at $50,000 per truck would require a $5 million capital expenditure. Based on the previous over-simplified example, a finance-leased fleet would require a $1.3 million expense per year for five years for $6.5 million. This fleet being leased rather than bought has $5 million of cash to invest in other projects, to earn the cost of finance-leasing with tax credits.

As a means of financing trucks, tractors, and trailers, leasing is receiving a great deal of analytical attention. This underscores the basic principle that should go into any investment decision, from obtaining a new truck to setting up a new plant: The vehicle should provide a satisfactory return on investment. How the investment is financed—leasing or buying—is a separate problem. The money saved by leasing should be reinvested or spent to generate income. It is a form of income that should earn enough to pay the extra expense of leasing. If it does not, outright ownership is the more desirable route. Leasing agreements have detailed penalties if you change before the end of the contract.

Finance leasing is a good way to get equipment into operation with a smaller amount of up-front money than direct purchase would require, but chances are that its cost in the long run will be higher than outright purchase. The lessor must purchase the vehicle, charge off capital expenses, cover overhead, and still make a profit. These items are, of course, passed along to the lessee.

On the advantage side, the leasing company may be able to obtain the vehicle at a savings through fleet-discount purchasing or possibly get financing at a favorable rate. But it is unlikely that these advantages will provide customers with a vehicle that costs less than an outright purchase.

Whatever the difference, it can be computed by comparing cash flows for the two situations and applying present value factors to the results; the added convenience of leasing may be worth the additional cost. There are intangible benefits: avoidance of the long-term operational commitment that is inherent in purchasing trucks, availability of a full-maintenance contract with a reliable company, guaranteed backup equipment, and budgeting of predictable expenses for the term of the lease. These are important to lessees.

Leased-truck fleets vary from very large to very small. Theoretically, there is a minimum point at which it is not economical for a company to perform its own maintenance. The exact size is in question. Estimates range from 10 trucks to 25 to vendor out the maintenance. Severity of use and geographic location are additional pertinent factors. In any event, the manager of a small fleet doing its own maintenance should be aware of all direct and indirect costs and compare those costs with contracts available through reputable full-service leasing companies. Another leasing advantage is the ease with which additional equipment can be obtained to meet seasonal or special requirements.

Additional Leasing Services

It is becoming more important to fleet managers that leasing companies provide other services to customers: safety training for company drivers, instructional courses on equipment, assistance with deciding on equipment specifications, and a whole array of financial data. The experience and expertise of the leasing company are especially valuable to the inexperienced fleet manager confronted with the need to obtain new equipment. The array of trucks and their various engine–transmission–rear axle combinations, accessory selections, and weight and strength classifications can be confusing.

If money freed up for leasing can earn, through tax credit and investment, the costs attributed to leasing, then it is a profitable way to go. Well-planned, this cash flow can allow company growth. To arrive at a decision to buy or lease, you must know the cost per vehicle in both instances. This would be the total direct and indirect costs divided by the number of vehicles in a class. Another way of measuring the dollar cost of vehicles is the cost per mile or per hour.

Many national leasing companies provide accounting and reporting services, handling state fuel tax and mileage charges. This kind of help is of particular value to small fleets that cross several states in the course of their operations because it relieves the fleets of having to remain up to date on confusing and ever-changing state regulations. The growing number of auxiliary services offered by truck leasing companies across the nation are figuring more in the lease-buy decision.

One caution: Do not assume that the lowest monthly lease payment is the best deal. There are various leasing arrangements available, and it is important to check with other lessees in your industry for their experiences before making a decision.

Finance Lease

The most popular, and generally most inexpensive, leasing option is the straight finance (capital) lease, in which a company leases trucks that the lessor has bought at the lowest possible price and keeps at locations convenient to its drivers. The lessor, who has a ready market, disposes of the vehicles at the completion of each lease cycle. A major feature of this plan is that the user in essence pays only the actual depreciation (the difference between the price the lessor paid for the vehicles and their selling price). There is no inflated estimate of depreciation included in the cost of leasing because the lessor does not guarantee a lid on depreciation costs. The lessee pays for all other costs, such as maintenance, fuel, and registration.

Full-Maintenance Lease

A popular option is the full-maintenance (operating) lease. Here, the lessor is responsible for all depreciation and maintenance costs, licensing and registration fees, physical damage insurance, interest costs, the lessor's administrative overhead, and specified miscellaneous fees. The lessor bills the lessee for these costs. The lessor pays for gas; oil between changes; liability and property damage insurance; accident and abuse repair expenses (up to deductible limits); sales, use, or personal property taxes; a mileage surcharge for any distance over the agreed-upon maximum; and, under certain circumstances, tire replacement. The lessor bills the lessee for these costs also.

Because of the additional services provided by the lessor, monthly payments for a full-maintenance operating lease are higher than for a finance lease. (However, there is a less expensive form of operating lease in which the lessor is not responsible for any operating costs.) The type of lease you select should depend not only on price but on how involved you are prepared to be in the details of managing and maintaining the fleet.

The leasing alternative becomes more enticing when senior management says, "No money will be budgeted for vehicle purchasing now." The response from the fleet manager should be to show the overall relative costs with vehicles not purchased versus vehicles leased.

49

There are two basic leasing alternatives:

1. Open end lease. The lessor owns the vehicle while the lessee is paying the principal and interest. At the conclusion of the lease term, the lessee owns the vehicle because it is paid in full.

2. Closed end lease. At the conclusion of the lease term, the leasing company owns the vehicle. In the case of a TRAC Leasing Agreement (Terminal Rental Adjustment Clause), a company could buy the vehicle with a balloon payment or an average price agreed on by both parties at the end of the lease period. Also, a higher up-front or back-end payment can be arranged during the lease period.

There are three basic lease alternatives (open or closed) from which you can choose:

1. The leasing company gives you the money and the lessor specifies, orders, maintains, and sells the vehicle—a finance lease.

2. The leasing company gives you the money, and it specifies, orders, takes delivery, gives you the vehicle, and maintains it—a full-maintenance lease.

3. The leasing company does all of the No. 2 alternative and supplies the driver—full operating and maintenance lease.

Each alternative gives you certain benefits for which you pay.

Another alternative to consider is for you to reimburse your employee for vehicle expenses.

Reimbursement Disadvantages

- No control of corporate image, vehicle age, or type of vehicle used.

- Reimbursement must be tailored to geographic locations.

- Vehicle maintenance and insurance are paid for at retail level.

- Limited tax deductions for employee.

- Higher depreciation paid (for higher mileage).

- Vehicle ownership is required of employee.

- Higher gas price is factored in driver's favor.

- The reimbursed employee vehicle could be different from the "standard vehicle" upon which reimbursement is based.

Leasing Advantages

- No corporate investment.

- Availability of multiple programs and financing sources.

- Ability to obtain lowest capitalized costs.

- Lessor expertise in vehicle recommendations.

- Ability to utilize multiple disposal options.

- Lessor provides record-keeping and collateral services.

- Reduced administration.

- Employees can purchase the vehicle at the end of the term.

- Age of the vehicle is controlled.

The financial condition of a company depends on income being greater than expenses. The more profit made, the more taxes paid.

The U.S. tax rules support keeping businesses healthy because businesses provide income to the workers who pay the federal government taxes. Each year, our government adjusts tax rules to keep our economy healthy. Although this is an oversimplification, the point is that economic strategies must be adjusted annually to be competitive and grow.

If you do not budget annually for vehicle replacements and the age of the fleet increases, there is a potential need for more manpower or vendors to maintain the aging fleet. This is counterproductive. If you use fleet replacement money to expand manufacturing capacity, older trucks could cost more to own and maintain because they get older. More vehicles are needed to handle increased production of goods.

Economic strategies of leasing offer an alternative cost plan to keep expenses in line because life-cycle costing supports an economic level of vehicle cost that is in the lessee's favor. Life-cycle costing defines the cost of old versus the cost of new. The objective is to support the most efficient level of vehicle cost. Too new, principal costs push total vehicle costs higher than proper aged vehicles. Too old, maintenance costs push total vehicle costs higher than proper aged vehicles.

If vehicles should be replaced to support fleet life-cycle costs, then do not postpone this action. Use economical alternatives that support rather than oppose this effort.

Budgeting Vehicle Replacement

In putting together a vehicle budget, the following cost information must be considered annually on a five to ten-year forecast. Use of historical information supports forecasting (Figure 2-6).

Include in the project:

1. Base annual costs - Principal.

2. Interest or depreciation costs.

3. Replaced vehicle costs—Additional cost for new vehicles above old vehicle cost.
    Old vehicle costs $1,000 per month, new vehicle costs $1,500 per month—Replacement cost is $500 per month more per vehicle until old vehicle is depreciated.

4. Deduct for vehicles matured or paid for.
    Vehicle that costs $1,000 per month now costs $50 per month, deduct $950 per month.

5. Reduce budget with income for vehicles sold.

6. Reduce from budget monies for vehicles retired and not replaced.

7. Add to budget for vehicles added to the rolls.

8. Add for inflation.

NUMBER OF VEHICLES PURCHASED

1995	1996	1997	1998	1999	2000	2001		Proposed Average Per Year
46	82	198	74	31	60	65	Cars	94
40	106	139	159	38	82	97	Pickups	150
19	10	30	14	-	1	22	Sm. Service Trucks	14
38	13	18	35	7	8	6	Miscellaneous Trucks	26
21	52	26	23	54	12	10	Heavy Crew Trucks	21
6	13	44	28	9	1	-	Heavy Work Trucks	12
170	276	455	333	139	164	200	Average Units	317/Yr.
							Per Year Leased	

PROPOSED ANNUAL VEHICLE ORDER

Total Inventory		Proposed Annual Vehicle Order		New Vehicle Cost	Dollars
376	Vehicles - 25%	94	Cars and Lights	$20,000 each	$1,880,000
599	Vehicles - 25%	150	Pickups and Light Vans	$25,000 each	3,750,000
68	Vehicles - 20%	14	Small Service Trucks	$70,000 each	980,000
265	Vehicles - 10%	26	Miscellaneous Trucks	$50,000 each	1,300,000
330	Vehicles - 10%	33	Heavy Crew/Work Trucks	$100,000 each	3,300,000
		317	Units/Yr.	Avg cost/unit =	$35,363 each
1,638 Vehicles			Replacement Cost	Estimated Annual	$11,210,000

Figure 2-6
Historical Purchasing Summary

9. Add for increase in vehicle prices.

10. Add for cost of money—Estimate interest rate increase.

11. Add a 5% contingency fee to the total of the previous 10 items.

This budgeting forecast projects for the company needs tomorrow. It allows management to include vehicle costs in operating revenue needs for tomorrow, and, should adjustments be needed, an understanding should be established that not buying vehicles for one year will move expenses into manpower and parts for next year and the following years. This is because older vehicles require more repair support and more operating support. This is self-perpetuating because there is a desirable age to replace a vehicle and to maximize cost efficiencies. Too old requires higher maintenance. Too new requires higher investment costs.

Vehicle Replacement Policy

In support of efficient fleet replacement programs, whether it be miles, hours, and/or years, a structured process should take place. For light vehicles (up to 10,000 lb), replacements should take place on economic evaluations.

For heavy vehicles (10,001 lb or greater), not only should economics be involved but also physical presence. The user, the maintenance people, and the replacement staff should visit each vehicle to be replaced and review its physical condition.

With the principle players face to face, a true vehicle evaluation will take place. Each can present and support reasons why, and a joint decision can be made to extend the life of the vehicle or replace it. At this point, dates can be established and expectations set for all three parties present.

Example: Policy—All 32,000-lb GVW vehicles will be replaced at 10 years.

The following vehicles qualify:

> Vehicle   #1234 - 12 years old
> #5678 - 11 years old
> #9012 - 10 years old
> #3456 - 9 years old (excessive maintenance to be needed)
> #7890 - 8 years old - severe accident, do not repair

When the appropriate people meet, the following is agreed:

Vehicle 1234: 12 years old—needs seat, paint, tires, and transmission repairs: will last 2 more years. Do not replace.

Vehicle 5678: Excessive rust: 100 miles per quart of add oil; needs 6 tires and transmission. Replace.

Vehicle 9012: Needs paint. Paint and extend.

Vehicle 3456:   Although 9 years old, replace. Premature failures; extensive maintenance needed.

Vehicle 7890:   Although 8 years old, accident damage excessive; do not repair. Replace it.

The more complex a vehicle is, the more investigations and evaluation are needed. With the concerned parties present, negotiation and compromise will determine the cost-effective action to take place.

Not Replacing Vehicles Can Be Expensive

Whether money for new vehicles is available is, of course, an important consideration, but if purchase money is budgeted annually, money could be saved in the long run. Withholding vehicle purchasing money will age a fleet and increase its maintenance costs. If company administration cuts funds for purchasing new vehicles, managers can offer some powerful arguments for a reconsideration.

These arguments are enhanced if records are available to project next year's costs. Compare the increased maintenance costs with new-vehicle costs, assuming the fleet is the proper size and is fully utilized and therefore will not require unusually large numbers of purchases. Any excess vehicles could be sold to generate additional new-vehicle purchase funds. Also, new vehicles can be introduced into a marginally utilized fleet so that two new vehicles will replace three old ones, further lowering overall fleet costs.

If these ideas can be effectively communicated to top management, the question of vehicle purchase money can be readdressed. The key factor, of course, is credible documentation, based on cost records to which you apply sound management principles.

During the service life of a truck, cost elements fluctuate—some increase, some remain level, and a few even decrease. Together, they represent the total cost of the unit, and this cost stream is the relevant factor in determining the economic life of the vehicle.

Concentration on only one or another cost factor, without considering the total, will most likely lead the equipment manager unwittingly to choose replacement equipment that produces lower costs in one or more elements but higher overall costs. This is a situation often forced on equipment managers by peculiarities of the budgeting process. For example, it is sometimes more difficult to obtain capital outlay dollars for vehicle replacement than operating dollars for maintenance. In reality, the amount of capital outlay may be far outstripped by the rapidly escalating costs of maintenance and downtime, not to mention costs that result from holding equipment beyond the economically optimum replacement point. Again, it shows that economic decisions normally cannot be made on the basis of a single cost element. Rather, they require that the equipment manager and the budget and/or finance officer sit down and view the total cost picture.

One means of dealing with this problem is to come to budget meetings armed with good data and analytical support for replacement proposals. Certain business considerations may dictate deviations from an economically optimum replacement program, but such decisions still should be made in light of objective analysis. This is preferable to the situation of many equipment managers who promote replacement plans using instinct and intuition rather than hard facts. Instinct and intuition may lead to proper conclusions, but these are easily dismissed during hard-nosed budget reviews.

Determination of Economic Lifetime

Assume your goal is to minimize the total costs associated with ownership of a particular piece of equipment. How do you determine the economic life for such a unit? Should it be replaced when costs start to rise or when average yearly costs are exceeded?

The answer, of course, is that new-vehicle costs must be compared with those of the current vehicle (per mile or per hour or cumulative cost per year). Today, higher fuel economy alone could justify purchasing a new vehicle instead of keeping less efficient, older units, even if those units are well-maintained.

Replacement Policies Under Budget Constraints

Up to this point, it has been implicit that given favorable cost projections, analytical methods, and replacement criteria, you would have sufficient funds available for purchasing new trucks when economically feasible. Although some may operate with few budget constraints, the vast majority of equipment managers must live with fund limitations that do not always permit them to replace vehicles at the economically optimum time. Realities of the budget process, legal limitations on borrowing, and administrative policies often leave equipment managers able to replace only a portion of those units due for replacement, if any at all.

Priorities must be set to determine which vehicles to replace with the available funds. The following method is one possible approach.

If a unit is due for replacement, the projected total costs of the current unit for the following year should be greater than the proposed replacement price. This difference is the estimated loss associated with holding the current vehicle beyond its economic point of replacement.

This priority ranking approach can be developed for an entire fleet, regardless of departmental assignment. If a given amount is appropriated for replacement of equipment, a *replacement priority rating list* could be constructed to act as a guide for replacement decisions. Priority ranking is intended to serve as a guide or a tool for decision making (Figures 2-7 and 2-8). It should in no way be construed as a substitute for the equipment manager's decision-making process. The vehicle assessment report shown in Figure 2-8 has a total of 68.8 points. The lower the total number, the greater the need to replace the vehicle because of poorer condition.

Many variables lead managers to decide to purchase new vehicles. *Limitations on the size* of the maintenance facilities may encourage a more vigorous replacement policy in the short run, because an older fleet will almost certainly require a greater maintenance commitment. A desire to *improve the environmental impact* of the fleet may suggest building a newer fleet having cleaner emissions. It may also require outfitting portions of the fleet with special fuel systems or powerplants to reduce pollution. A management desire for a *"shiny new fleet image"* may dictate shorter replacement cycles. *Bureaucratic notions, expectations, and rivalries* also affect replacement decisions. In the long run, economic conditions within the boundaries of an agency influence the manager's decisions, shortages of revenue, and constrained borrowing capacity set the limits on any and all replacement decisions. These considerations, and others, influence decision making and dictate which units should be traded and when those units will be replaced.

VEHICLE NO.:		VEHICLE MAKE:		MODEL:		YEAR:	
Rating Legend:	5 = Excellent	4 = Very Good	3 = Good	2 = Average		1 = Poor	

Inspection Results

Description	Awarded Rate		Multiplier		Actual Rate	Max. Rate	Factor - Score	
Section 01 Body/Interior			TOTAL MAX. POINTS 15					10.8
Rust	3	×	20	=	60	100		
Condition	3	×	20	=	60	100		
Accident Damage	4	×	40	=	160	200		
Glass	5	×	10	=	50	50		
General Interior Condition	5	×	10	=	30	50	360	72%
				Total	360	500	500	
Section 02 Tires			TOTAL MAX. POINTS 15					9.0
Tread Wear	3	×	60	=	180	300		
Sidewall Cracking	3	×	40	=	120	200	300	60%
				Total	300	500	500	
Section 03 Driveability			TOTAL MAX. POINTS 15					9.0
Road Test	3	×	100	=	300	500	300	60%
				Total	300	500	500	
Section 04 Brake System			TOTAL MAX. POINTS 15					15.0
Stop Ability	5	×	60	=	300	300		
Emergency	5	×	40	=	200	200	500	100%
				Total	500	500	500	
Section 05 Steering/Suspension			TOTAL MAX. POINTS 10					8.0
Loose	4	×	20	=	80	100		
Vibration	4	×	20	=	80	100		
Pulling	4	×	20	=	80	100		
Standing Appearance (level/square to road)	4	×	40	=	160	200	400	80%
				Total	400	500	500	
Section 06 Electrical			TOTAL MAX. POINTS 10					10.0
Once Over Safety Items	5	×	100	=	500	500	500	100%
				Total	500	500	500	
Section 07 Driveline			TOTAL MAX. POINTS 10					10.0
Noise	5	×	20	=	100	100		
Vibration	5	×	20	=	100	100		
Leaks	5	×	20	=	100	100		
Shifting	5	×	40	=	200	200	500	100%
				Total	500	500	500	
Section 08 Engine			TOTAL MAX. POINTS 10					8.0
Leaks	4	×	30	=	120	150		
Rough Running	4	×	40	=	160	200		
Noise	4	×	30	=	120	150	400	80%
				Total	400	500	500	
	TOTAL 92		TOTAL RATE AWARDED		3260		TOTAL SCORE	79.80

Total Score Legend:
3200–4000 Excellent
2400–3200 Very Good
1600–2400 Good
800–1600 Average
0–800 Poor

Awarded Score Legend:
88–110 Excellent
66–88 Very Good
44–66 Good
22–44 Average
0–22 Poor

Figure 2-7
Vehicle Assessment Report
Light-Duty Vehicle

VEHICLE NO:	VEHICLE MAKE:		MODEL:		YEAR:	
Rating Legend:	5 = Excellent	4 = Very Good	3 = Good	2 = Average	1 = Poor	
Description	Awarded Rate	Multiplier	Actual Rate	Max. Rate	Factor-Score	
Section 01 Chassis		TOTAL MAX. POINTS = 15				10.8
Rust and Corrosion	3 ×	20 =	60	100		
Condition	3 ×	20 =	60	100		
Accident Damage	4 ×	40 =	160	200		
Glass	5 ×	10 =	50	50		
Interior	3 ×	10 =	30	50	360	72%
		Total	360	500	500	
Section 02 Tires		TOTAL MAX. POINTS = 15				9.0
Tread Wear	3 ×	60 =	180	300		
Sidewall Condition	3 ×	40 =	120	200	300	60%
		Total	300	500	500	
Section 03 Body Mounted Equipmt.		TOTAL MAX. POINTS = 15				9.0
Dump Body Exterior	3 ×	20 =	60	100		
Dump Body Interior	3 ×	20 =	60	100		
Tail Gate	3 ×	20 =	60	100		
Extension Boards	3 ×	20 =	60	100		
Lift Cylinder	3 ×	20 =	60	100	300	60%
		Total	300	500	500	
Section 04 Brake System		TOTAL MAX. POINTS = 15				12.0
Service Brakes	4 ×	60 =	240	300		
Emergency Brakes	4 ×	40 =	160	200	400	80%
		Total	400	500	500	
Section 05 Steering/Suspension		TOTAL MAX. POINTS = 15				12.0
Looseness	4 ×	20 =	80	100		
Vibration	4 ×	20 =	80	100		
Pulling	4 ×	20 =	80	100		
Parallel to Ground	4 ×	40 =	160	200	400	80%
		Total	400	500	500	
Section 06 Engine and Driveline		TOTAL MAX. POINTS = 25				16.0
Leaks	2 ×	20 =	40	100		
Vibration	2 ×	20 =	40	100		
Noise	3 ×	20 =	60	100		
Shifting	4 ×	20 =	80	100		
Rough Running	5 ×	20 =	100	100	320	64%
		Total	320	500	500	
TOTAL 79		TOTAL RATE AWARDED	2080	TOTAL SCORE		68.8

Total Score Legend:
2400–3000 Excellent
1800–2400 Very Good
1200–1800 Good
600–1200 Average
0–600 Poor

Awarded Score Legend:
92–115 Excellent
69–92 Very Good
46–69 Good
23–46 Average
0–23 Poor

Figure 2-8
Vehicle Assessment Report
Heavy-Duty Vehicle

Chapter 3

Vehicle Life-Cycle Costing
⁜

Cost Analysis

At some time, we must evaluate the productivity of what we are doing and compare it to what we could be doing more efficiently as a result of technology, such as tool upgrades, work method changes, and market changes.

An example would be our life-cycle cost with our equipment being sold at, say, eight or nine years with a four-year average age. Today, should there be a glut of used equipment at two- and four-year age averages, our resale revenue is reduced, and we must adjust our vehicle costs for our present life cycle. However, does this trend help to lower our cost and support our reliability?

From where do we obtain the numbers and information necessary for us to compare accurately?

Regulatory agencies require that we have accounting figures for those agencies to review and analyze. From the figures of the accounting department, we can examine our capital and operating costs. Another source of good information is our management information systems, where we can obtain class maintenance data.

Another source of viable and related information is the vehicle chassis electronic control unit. We obtain information on miles, hours, gallons, temperatures, speeds, and idle to compare to our schedules. This educates us so we can understand, synergize, and formulate improvement plans.

With this information, we can cost-analyze our operations and determine what issues we have.

It is important to recognize and understand the issues so we can communicate to the line-level people the various alternatives for improved goals and objectives. This also allows us to partner with them for improvements to achieve better safety and efficiency.

Another technique is to sequence what we are doing differently to recover better work standards without sacrificing the safety, quality, and reliability of our delivered tasks. It all starts by examining parts of the whole and adjusting the process to be more efficient. Deming and Crosby would be proud. Someone is listening.

For example, chassis costs and mounted equipment costs have different footprints.

Chassis manufacturers sell large volumes and have quality engineers to sort out process issues. They work out more of the bugs from their designs because more end-user information is available due to advancements in management information systems and the volume they sell.

Mounted equipment and bodies are produced in peak-and-valley order sequences that make it difficult to pay attention to reliable and meaningful indices. To explain, although standardization exists, many applications are different. However, by using the cost analysis and information that fleet managers have kept for the past 30 years, the mounted equipment people can determine end-user needs and make improvements to their products.

Thus, cost analysis has a solid root in fleet management because it helps you in identifying root causes and in developing cost alternatives for communicating to work study groups/teams to develop better ways of doing things.

Cost analysis requires accurate information that has standardized terminology (e.g., miles per gallon). We have differences with accounting numbers and fleet management numbers; therefore, we must review all inputs, find the accurate source, and communicate it.

Cost analysis requires interpretation of the numbers. Numbers are objective information, but a person or persons with savvy must interpret the application strategy and post-estimate numbers for the proposed alternative strategy or strategies. The most important part of cost analysis is the use of historical information that was tested and applied to alternatives and the inclusion of real-world people in the formulative process.

Communicate and cooperate to obtain the buy-in. To propose alternatives without the "doers" is a void in the quality process.

Plan our work and work the plan. We must cooperate internally to be competitive as a company.

It's All About Location

If we put four identical vehicles into inner-city, city, suburban, and rural applications, each will yield different maintenance and operating costs.

Inner-city costs in those areas will be higher in cost and cost per mile than in the city, suburban, and rural areas because of the pounding the vehicle takes resulting from vibration, moisture, corrosion, and accidents.

Tires experience tread and sidewall damage from potholes and curbs. Brakes become heated with continuous stop and go. On the other hand, the density of an inner-city environment can give fuel cost relief.

How do fleet professionals address the variance in costs?

We benchmark our costs by grouping similar environments for compatible cost information analysis and defining life-cycle periods that make economical sense.

We will find cluster costs; that is, 70% of our total chassis costs will be concentrated in preventive maintenance, tires, brakes, suspension, electrical, HVAC, engine, and steering. The percentage in each of these maintenance repair codes will vary, based on the application-specific environment.

In an inner-city application, we might exhibit more rapid wear cycle costs, thereby shortening our expected life-cycle periods. Many fleets strategize that by assigning a vehicle to low cost applications (rural and suburban) for a good portion of its life and when the vehicle starts to incur increased maintenance and operating costs, reassign it to the city and then to the inner-city location. This lowers vehicle mileage but increases vehicle costs significantly. Paraphrased, this means we cannot control severe cost applications environment.

Thus, we put the oldest and poorest vehicles in this environment, and we will tear up an old vehicle rather than a new vehicle.

To address this from another perspective, if we spec a vehicle properly, we will obtain the best possible life cycle from it rather than take a standardized vehicle, use it for several applications, then finish its usage in the worst unspecified environment. It sounds as if we are exacerbating our severe-duty application costs.

Looking at our extreme-duty applications, what specs can we upgrade to extend vehicle life? Suspensions can be upgraded to deal with severe on-/off-road application using air support springs and shock absorbers to dampen suspension ranges. It costs more to spec the unit, but that is what the application environment is demanding from our vehicles.

Brakes require frequent usage resulting in higher operating temperatures. Brake shoe/pad foundation material could be installed with harder cerametallic compounds to work better in a high-temperature application. Our component costs will be higher, but our maintenance costs in this application will be lower than moving an older work vehicle into this environment when that vehicle has not been specified for this application. We will have shortened reliability cycles and the higher costs that result in more severe cost spikes.

The preventative maintenance inspection program should be intensified for severe-duty inspection applications. For example, expand system inspections from belts and hoists and hold-downs to bracket mounts and critical bolt torquing to expand into areas that will develop into maintenance items because of vibration environments.

Electrical component mounting connections that are affected by moisture, debris, road trash, close passageways, vandalism, and other cultural effects characterized by inner-city locations also should be inspected. During the inspection process, road debris can be removed and wires retied, rerouted, and repaired to reduce failures. Our staff should be trained in severe-duty inspection techniques so they are aware of our expectations.

Lubrication inspectors bring to bear some modified techniques with steering systems. Let us take kingpins, for example. We inspect the play and clearances and then lubricate, noting how much grease is taken before the clean grease starts to push out. Two or three pumps start to indicate kingpin bushing wear.

Although this condition is still safe, we should prepare for a component replacement and estimate a close future date for service. We also should obtain the part in stock for use at the next periodic maintenance (PM) inspection and plan to do this work and notify the user that the next PM will be longer than usual because of the deferred scheduled maintenance planned.

Now we can plan for a scheduled component repair. These strategies allow us to be proactive to meet our challenges in a predictable or scheduled way. After all, we are dealing with consumables that must be measured and, when worn, must be replaced before they break. It is cheaper to replace items prior to failure rather than after failure.

When we assign vehicles, how we use them results in variable costs. Yes, we have a class average, but the extremes must be addressed in terms of useful life. Although we have a targeted life, we can extend or shorten it based on its resultant cost from its assigned location. In fact, we can predict these costs in advance, based on historical information per vehicle and its trending.

Fleet managers have much in common with other asset managers. What are the three most important elements to consider for successful investment of funds? Location, location, and location.

General Information

When do you buy a vehicle? When you have money. Do you lease or own? That depends on the amount of money you have. If you're cash rich, you own. If you have poor cash flow, you use someone else's money in leasing. Leasing allows you to buy new, updated vehicles with engineering changes to take advantage of productivity demands of your operations and also lower operating and maintenance costs due to new technology. Leasing has rules to protect the investor, so we are held to these rules. Borrowing money gives us some flexibility, as does cash. We must keep this in mind. Vehicle maintenance costs of old vehicles change based on use, proper or improper application, fleet mix, density, and operational changes.

The buy, rebuild, or stand-pat analysis must be made annually to take into consideration your ever-changing financial, operating, and maintenance variations.

First here is a look at the pattern of vehicle costs.

To maintain the lowest cost and maximum vehicle availability for top utilization, older vehicles should be replaced when the cost to operate and maintain them is higher than the cost of a new vehicle or when technical obsolescence becomes apparent. This is a basic concept of life-cycle costing and good business common sense.

Ownership costs are the principal and interest costs incurred on a monthly basis until the vehicle is paid off. Another example would be the ownership cost depreciated (reduced on a fixed time period, such as five years) until its full value is reached.

Operating and maintenance costs are a familiar item. Operating costs are, for example, fuel, taxes, and registration. Maintenance costs are labor and parts, basically, no matter in what industry, common carriage, private carriage, municipality, utility, and governmental areas. The ratio of parts to labor dollars as illustrated in Figure 3-1 through Figure 3-3 shows an average of a 1:1 dollar ratio. Figure 3-1 supports this fact with ratio breakdowns per component. Although the average ratio is 1.4–2.7, the dollar volume in Figure 3-3 shows a 1:1 average.

In Figure 3-2, note that on over-the-road vehicles, normal maintenance parts-to-labor ratios are 44% parts, 56% labor. Young vehicles tend to be labor intensive, when the average age (consumable and component change outs) 50:50 than when being rebuilt or heavy repairs changed out the ratios increases to 66:75 parts, 34:25 labor.

Figure 3-3 shows 60% of the total repair costs are in seven component areas that also have a 1:1, 50:50 parts-to-labor ratio and does not include any rebuilding costs.

These costs can be tracked in a manual or an automated fashion to accumulate life costs. Ownership and operating costs are the most predictable. Maintenance is a less predictable cost, showing peaks and valleys as illustrated in Figures 3-4 and 3-5. As a vehicle ages, additional costs are incurred; by measuring these additional costs, future performance can be predicted based on this historical information. Note Figure 3-4 through Figure 3-8, which Figure 3-4 shows cumulative life costs by component for parts and labor, and Figures 3-5 to 3-8, which show component costs per year, or in this case,

Chassis	Parts	Labor	Mounted Equipment	Parts	Labor
Brakes Replace	2	1	Hydraulic Pump	4	1
Brakes Repair	1	2	Hydraulic Valve	1	2
Frame	1	3	Hydraulic Cylinder	1	1
Steering	1	1	Hydraulic Fittings	1	3
Axle	1	1	Bucket	1	5
Suspension	1	1	Boom	1	5
Wheel and Rims	1	2	Boom Access	1	5
Rear Axle	2	1	Turret	1	6
Clutch Replace	1	1			
Clutch Repair	1	5	Mounted Equip. Total	11	28
Drive Shaft	1	2		1.4	3.5
PTO	1	1		30%	70%
Transmission RXR	1	1			
Transmission Repair	1	2			
Cooling	1	1			
Exhaust	1	2			
Fuel	1	2			
Engine Replace	2	1			
Engine Repair	1	2			
Charging	1	1			
Starting	1	2			
Lights	1	3			
Dash—Interior	1	3			
Tires	8	1			
Chassis Total	34	42			
	1.4	1.8			
	45%	55%		Parts	Labor
			Summary Chassis	1.4	1.8
			Mounted Equipment	1.4	3.5
			Average	1.4	2.7
				38%	62%

Figure 3-1
Ratio of Parts to Labor Dollars

15,000-mi summaries. Each 15,000-mi period can be examined, noting the high cost areas. These areas can be identified as scheduled or unscheduled costs. Scheduled component costs are desirable because they are lower than an equivalent unscheduled component cost. Brake replacement is scheduled; brake repair is unscheduled. As inherent costs are tracked in this class in the operating environment, future costs for new models for this class can be predicted based on historical data. By modifying your vehicle specifications, you can attempt to limit unnecessary future expenses. An example is to spec higher-capacity brakes to lower brake service costs, shock-proof lighting

			Fleet Maintenance Cost per Mile Mileage Increment				
From	0	100,001	200,001	300,001	400,001	500,001	0
To	100,000	200,000	300,000	400,000	500,000	600,000	600,000
			Average Cost per Mile (in $)				
Parts	0.0109	0.0237	0.0305	0.0369	0.0309	0.0389	0.0257
	(28%)	(46%)	(47%)	(44%)	(48%)	(50%)	(44%)
Labor	0.0281	0.0281	0.0349	0.0466	0.0333	0.0387	0.0330
	(72%)	(54%)	(53%)	(50%)	(52%)	(50%)	(56%)
Total	0.0390	0.0518	0.0654	0.0835	0.0642	0.0776	0.0587
			Average Miles per Repair Order				
	5,237	3,536	2,804	2,466	3,140	3,613	3,316
	19.1	28.0	35.7	40.6	22.6	27.7	173.7
			Part/Labor Ratio: Parts 44%, Labor 56%				

Figure 3-2
Examples of Ratio of Parts to Labor
(Source: National Aftermarket Data Exchange)

Type Description	No. Vehicles	No. R/O	Total DOL	Comm DOL	Parts DOL	Labor DOL
Tires	50	659	110,803	3,021	94,875	12,906
Brakes Replace	41	162	57,943	33	31,324	20,585
Brakes Repair	50	563	40,903	–	15,907	24,997
Power Replace	28	43	39,679	–	26,968	12,693
Power Repair	49	336	30,114	42	10,065	20,007
Preventive Maintenance	150	1,501	68,124	–	3,410	64,714
Suspension	44	302	46,865	265	22,978	23,621
Lighting	53	1,616	40,129	–	11,671	28,458
Steering	44	319	32,273	106	14,326	17,842
Total			457,347		231,524	225,823
			60%		50.62%	49.48%
Total			$759,548		$363,978	$391,282

Figure 3-3
Maintenance-Component Life-Cycle Costs of Vehicle Class

600 CS Maintenance Analysis By Class			Sheet 1			Class 32 All All
			***Maintenance			
Repair Num/Type Desc	Num Veh	Num R/O	Total Dol	Comm Dol	Parts Dol	Labor Dol
PMA	52	755	23,418		8,713	14,705
PMB	51	564	12,458		300	12,158
PMC	46	181	16,471		381	16,090
PMD	1	1	25			25
PME						
PMF						
PMG						
Sub Total			52,373		9,395	42,978
111 Axle Non D	25	33	5,550	125	2,756	2,669
113 Brakes Replace	41	162	57,943	33	37,324	20,585
213 Brakes Repair	50	563	40,903		15,907	24,997
114 Frame	31	81	6,764		1,430	5,335
115 Steering	44	319	32,273	106	14,326	17,842
116 Suspension	44	302	46,865	265	22,978	23,621
118 Wh,Rim,H&B	42	231	16,275	42	5,459	10,774
121 Axle Dr F						
122 Axle Dr R	35	107	12,555		6,039	6,516
123 Clutch Replace	26	65	21,967		13,554	8,413
223 Clutch Repair	31	159	3,366		572	2,795
124 Dr Shafts	35	86	4,762		1,825	2,937
125 P T C	43	191	8,470		3,431	5,039
126 Trans Replace	21	20	6,578		3,228	3,350
226 Trans Repair	41	147	18,914		7,444	11,470
128 Trans Aux	1	1	12			12
141 Air Intake	33	101	3,072		1,190	1,882
142 Cooling	48	281	11,765	38	5,540	6,187
143 Exhaust	42	249	9,698		3,681	6,018
144 Fuel Sys	53	643	21,353		7,229	14,124
145 Power Replace	28	43	39,679		26,986	12,693
245 Power Repair	49	336	30,114	42	10,065	20,007
131 Charge Sys	46	288	11,950		6,503	5,447
132 Crnk & Bat	40	154	6,692		2,980	3,712
133 Ignition	39	528	24,242		8,799	15,443
134 Lighting	53	1,616	40,129		11,671	28,458
103 Ins & Gage	46	149	4,969		1,258	3,711
102 Cab Doors	49	321	9,518	189	3,982	5,347
Sub Total			496,376	840	226,156	269,380
701 Hyd Pump	36	77	18,912		15,470	3,442
702 Hyd Motor	25	26	1,258		270	988
703 Hyd Valves	48	193	14,524		4,691	9,833
704 Hyd Cylin	34	75	6,359		3,159	3,200
705 Hyd Fittings	48	304	18,059		4,297	13,761
706 Bucket	39	118	6,346		988	5,359
707 Boom	50	348	24,011	427	3,545	20,039
708 Boom Accessories	29	51	4,076		622	3,454
709 Turret	35	89	6,343		510	5,833
710 Outrigger	3	1	83			83
711 Winch	4	3	26			26
117 Tires	50	659	110,803	3,021	94,876	12,906
Sub Total			210,799	3,448	128,427	78,924
All Other Repair Types Total			759,548	4,289	363,978	391,282

Figure 3-4
Total Vehicle Lifetime Component Cost Summary

600 CS Maintenance Analysis By Class		Sheet 2	Class 32 All All	
	Year 1 Zero To 15,000		**Year 2** 15,001 To 30,000	
RepairNum/Type Desc	Num Veh	Num R/O	Num Veh	Num R/O
PMA	51	4,543	44	4,118
PMB	49	2,287	41	2,238
PMC	30	2,146	31	2,805
PMD	1	25		
PME				
PMF				
PMG				
Sub Total	52	9,001	44	9,162
111 Axle Non D	3	288	2	175
113 Brakes Replace	13	1,522	24	4,418
213 Brakes Repair	43	4,155	45	5,490
114 Frame	13	684	12	606
115 Steering	25	1,310	35	2,829
116 Suspension	25	2,404	28	3,685
118 Wh,Rim,H&B	21	1,300	26	1,578
121 Axle Dr F				
122 Axle Dr R	16	406	11	2,144
123 Clutch Replace	5	1,009	9	3,435
223 Clutch Repair	14	397	17	287
124 Dr Shafts	8	332	13	1,372
125 P T C	25	1,480	20	1,891
126 Trans Replace			2	1,471
226 Trans Repair	17	448	18	3,045
128 Trans Aux	1	12		
141 Air Intake	13	207	12	329
142 Cooling	30	1,259	31	1,860
143 Exhaust	22	1,177	28	1,116
144 Fuel Sys	44	3,601	42	3,187
145 Power Replace	4	3,628	5	107
245 Power Repair	39	2,050	24	3,778
131 Charge Sys	22	1,552	29	2,272
132 Crnk & Bat	20	1,014	20	1,115
133 Ignition	25	1,747	32	3,994
134 Lighting	53	8,666	48	7,828
103 Ins & Gage	21	669	25	884
102 Cab Doors	41	1,588	27	1,167
Sub Total	53	42,907	48	60,062
701 Hyd Pump	13	2,167	11	4,893
702 Hyd Motor	7	183	6	198
703 Hyd Valves	25	2,218	25	3,408
704 Hyd Cylin	10	630	14	872
705 Hyd Fittings	36	3,259	34	3,178
706 Bucket	22	970	15	565
707 Boom	43	3,431	35	4,251
708 Boom Accessories	10	316	16	999
709 Turret	10	702	11	1,219
710 Outrigger			1	26
711 Winch	1	5	2	16
117 Tires	36	8,843	46	23,722
Sub Total	52	22,724	48	43,347
All Other Repair Types Total	53	74,632	48	12,571
Cents Per Mile		9.700		16.500
Curr Veh Per Inter	5		11	
Total Miles Accum.		768,577		681,159
Miles Current Per		4,201		18,418

Figure 3-5
Costs for Year 1 and Year 2

600 CS Maintenance Analysis By Class		Sheet 2	Class 32 All All	
RepairNum/Type Desc	Year 3 30,001 To 45,000		Year 4 45,001 To 60,000	
	Num Veh	Num R/O	Num Veh	Num R/O
PMA	37	3,743	31	3,215
PMB	35	2,060	30	1,849
PMC	30	3,004	27	2,580
PMD				
PME				
PMF				
PMG				
Sub Total	38	8,808	31	7,744
111 Axle Non D	7	380	3	199
113 Brakes Replace	26	10,874	21	9,195
213 Brakes Repair	37	8,039	30	10,014
114 Frame	13	1,958	12	805
115 Steering	33	4,611	21	7,250
116 Suspension	29	9,856	26	5,647
118 Wh,Rim,H&B	27	3,544	21	1,728
121 Axle Dr F				
122 Axle Dr R	21	2,879	12	967
123 Clutch Replace	11	3,574	12	3,908
223 Clutch Repair	21	703	20	733
124 Dr Shafts	13	697	9	619
125 P T C	19	1,055	18	1,279
126 Trans Replace	3	98	7	2,485
226 Trans Repair	22	3,200	18	2,685
128 Trans Aux				
141 Air Intake	23	778	17	476
142 Cooling	26	1,895	23	1,637
143 Exhaust	26	1,553	20	979
144 Fuel Sys	32	2,925	29	3,060
145 Power Replace	5	6,071	5	6,921
245 Power Repair	26	7,063	24	6,351
131 Charge Sys	27	1,524	25	2,386
132 Crnk & Bat	20	1,014	17	891
133 Ignition	32	3,936	28	4,519
134 Lighting	38	6,201	32	4,915
103 Ins & Gage	21	798	14	902
102 Cab Doors	26	1,722	21	1,399
Sub Total	40	86,948	33	81,948
701 Hyd Pump	12	2,929	14	2,492
702 Hyd Motor	6	191	4	126
703 Hyd Valves	20	1,951	21	1,961
704 Hyd Cylin	14	1,469	12	1,239
705 Hyd Fittings	22	2,506	24	1,754
706 Bucket	17	1,462	12	1,298
707 Boom	30	5,906	27	2,263
708 Boom Accessories	12	464	6	461
709 Turret	19	2,110	14	798
710 Outrigger				
711 Winch				
117 Tires	36	21,115	33	14,582
Sub Total	39	40,104	33	26,974
All Other Repair Types Total	40	35,859	33	16,667
Cents Per Mile		25.200		24.600
Curr Veh Per Inter	4		4	
Total Miles Accum		539,814		474,289
Miles Current Per		15,001		4,777

Figure 3-6
Costs for Year 3 and Year 4

71

600 CS Maintenance Analysis By Class	Sheet 2		Class 32 All All	
	Year 5 60,001 To 75,000		Year 6 75,001 To 90,000	
RepairNum/Type Desc	Num Veh	Num R/O	Num Veh	Num R/O
PMA	28	2,355	27	1,850
PMB	28	1,359	26	1,068
PMC	23	2,473	14	1,465
PMD				
PME				
PMF				
PMG				
Sub Total	30	6,186	27	4,383
111 Axle Non D	8	471	7	1,653
113 Brakes Replace	24	12,503	15	8,752
213 Brakes Repair	28	3,393	21	2,430
114 Frame	12	1,444	13	976
115 Steering	27	4,328	20	5,163
116 Suspension	23	5,192	21	7,807
118 Wh,Rim,H&B	24	2,788	19	1,812
121 Axle Dr F				
122 Axle Dr R	8	508	10	1,727
123 Clutch Replace	11	3,335	8	2,518
223 Clutch Repair	13	392	11	220
124 Dr Shafts	13	539	9	350
125 P T C	16	638	13	981
126 Trans Replace	5	474	5	1,471
226 Trans Repair	18	4,587	9	761
128 Trans Aux				
141 Air Intake	14	544	13	382
142 Cooling	24	1,428	19	1,438
143 Exhaust	17	1,369	17	1,118
144 Fuel Sys	23	2,573	25	2,641
145 Power Replace	5	5,591	6	4,040
245 Power Repair	20	2,442	24	3,955
131 Charge Sys	23	1,468	14	702
132 Crnk & Bat	17	1,027	16	622
133 Ignition	25	2,879	24	2,809
134 Lighting	20	4,438	27	4,299
103 Ins & Gage	15	515	13	442
102 Cab Doors	19	1,032	18	1,288
Sub Total	30	65,897	28	60,358
701 Hyd Pump	7	2,052	7	1,829
702 Hyd Motor	4	119	4	83
703 Hyd Valves	20	1,594	12	709
704 Hyd Cylin	8	210	5	376
705 Hyd Fittings	23	4,292	13	605
706 Bucket	11	422	8	516
707 Boom	21	2,044	14	1,321
708 Boom Accessories	6	981	4	62
709 Turret	10	786	5	150
710 Outrigger				
711 Winch	1	4		
117 Tires	29	12,721	27	13,074
Sub Total	30	25,226	28	18,724
All Other Repair Types Total	30	97,310	29	83,465
Cents Per Mile			23.300	22.500
Curr Veh Per Inter	2		8	
Total Miles Accum			418,383	371,180
Miles Current Per			1,415	8,635

Figure 3-7
Costs for Year 5 and Year 6

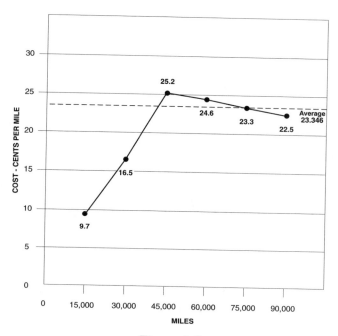

Figure 3-8
Maintenance Cost per 15,000-Mi Annual Interval
Total of 6 Years

and/or LED (light emitting diodes) to reduce lighting costs, and higher-capacity alternators to capture lower costs on electrical starting and charging system service life. If you succeed with one brand rather than another, you can incorporate the most cost-effective brand in your specifications, taking care to add or equivalent after each brand specified to encourage competition.

This cost identification is dependent on meticulous data input so that you can extract statistical data appropriately formulated for clear evaluation.

The objective here is to replace an old vehicle immediately prior to incurring a significant cost and after amortizing a large investment. By grouping the same vehicles together, you can set up a class average and compare your units to the class average. If a vehicle is above the class average, you can look at the historical cost trends for the vehicle, compare the average for the class and the average for the vehicle, and fix an economical replacement time.

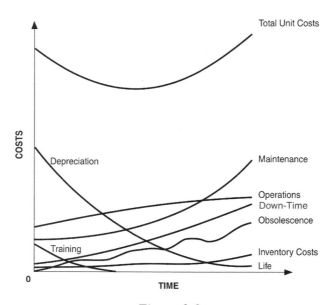

Figure 3-9
Fleet - Equipment Cost Concept
(Courtesy Byrd, Tallamy, McDonald, and Lewis, Falls Church, VA)

You then can write a policy on the proper economical replacement of a vehicle or class of vehicles. Age affects the maintenance cost information for a vehicle (Figure 3-9).

Total unit costs are impacted by each of the following costs and the rate of which they rise alone and together:

1. **Depreciation**

   Depreciation is a fixed cost, different in each business application and a product of a lease or own objective. It is because of the downward trend of this cost that if the vehicle is efficiently maintained, its useful life lengthens. A new vehicle reflects higher principal and interest due to the higher purchase price of new. Today, with the higher cost of money, short or long term, and the difficult economic climate affecting business growth and profitability, you have a strong case for extended vehicle life. Resale

```
Capitalization
        Gross Revenue                $100,000,000
        Operating Expense              90,000,000
        Profit                       $ 10,000,000
        Capital Tax 34%                 3,400,000
        Net Profit                   $  6,600,000   Alternative #1

Purchase $20,000,000 in Vehicles
5-Year Depreciation $4,000,000 per year
        Gross Revenue                $100,000,000
        Operating Expense              90,000,000
        Profit                       $ 10,000,000
        Vehicle Depreciation            4,000,000
        Net                          $  6,000,000
        Capital Tax 34%                 2,040,000
                                     $  3,960,000
                                        4,000,000
                                     $  7,960,000   Alternative #2
*Alternative #2 is $1,360,000 more cost-effective each year,
          $6,800,000 more cost-effective for five years.
```

Figure 3-10
Benefit from Depreciation

values due to supply and demand of new vehicles impact depreciation rates. You should measure these each year to include them in your analyses.

A good example of the textbook economic benefit from depreciation is the corporate tax paid by a company on its profits. Figure 3-10 looks at a gross revenue of $100 million netting $10 million profit. The government would say 34% of that amount must be paid in taxes, which equals $3.4 million, netting $6.6 million in profit to be used to pay stockholders, develop the business, and pay as bonuses.

If we purchase $20 million worth of vehicles and equipment and allow them to depreciate over 5 years in a straight-line fashion, we would pay 34% of $6 million per year rather than $10 million, thus netting $7.96 million per year in profit.

An extra $1.36 million per year over 5 years yields $6.8 million more in profit, which would mean that instead of $20 million paid for vehicles and equipment, we really paid $13.2 million. This rewards the company for stimulating the economy.

2. **Operations**

   This cost tends to gradually increase due to increasing fuel and related costs.

3. **Maintenance**

   Maintenance (Figure 3-11) tends to increase as the vehicle ages, and it represents the largest impact on the diagnostic cost benefits. If you hold this escalating cost down with effective scheduled maintenance (Preventive

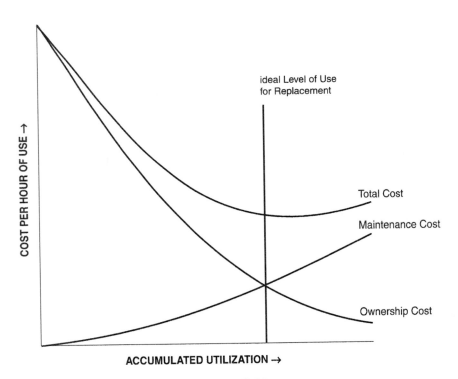

Figure 3-11
Relationships Between Equipment Ownership
and Maintenance Costs
(Courtesy Byrd, Tallamy, McDonald, and Lewis, Church Falls, VA)

Maintenance Inspection [PMI]—Effective driver write-up control and timely component replacement prior to failure), as long as obsolescence and technical life do not surface, you can keep vehicles longer before cost-effective replacement.

4. Down-Time

When a vehicle is not available for service, it has a per hour, per diem, or per mile cost. This cost should be indexed and unacceptable variations highlighted.

5. Obsolescence

Obsolescence is a function of work needs—If the crew can't use the vehicle because the vehicle is not configured for efficient task use, remove this vehicle from the fleet and replace it with a vehicle spec'd out to complement productive work methods. Use the resale revenue received or the productivity gains to offset the purchase price of the new vehicle.

6. Inventory Costs

Managing the parts and supply operation impacts a small part of your budget cost. Managing this area profitably requires big-ticket-item control primarily to support the repair function. Because this area represents a small part of your life-cycle costs and its predictability is relatively constant with small variations, it requires the least amount of your time and attention in relation to the whole life-cycle process.

7. Life Costs

Life costs are fixed and variable expenses over time and mileage.

When you have a fix on fleet age by type of vehicle (Figure 3-12) and class (Figure 3-13), you can move into a brief analysis of a new vehicle cost compared to an old vehicle cost (Figure 3-14 through Figure 3-16).

Each cost is listed on Figure 3-15 to prepare for an economic analysis—Figure 3-16 and Figure 3-17. The cost of money and tax write-ups have been omitted because this is the controller's area of input. Do the controllers want to identify vehicle cost to the capital or expense side? This depends on their corporate strategy and the availability of cash and the company financial structure. The tax rules change here quite often, and you must be sensitive to

Vehicle Category	Number of Vehicles	Average Age (Years)
Passenger Vehicles	105	4.6
Trucks	1,242	5.3
Pickups	457	4.1
Vans	267	4.7
Platforms	37	7.3
Aerial Devices	244	5.8
Line Trucks	163	8.5
Various Utility	67	5.9
Highway Tractors	7	6.1
Power Equipment	61	10.7
Trailers	515	15.2
Equipment	68	6.7
Transmission Distribution	320	6.1
Mobile Subs & Transformers	18	15.5
Miscellaneous Utility	19	14.5
TOTAL	1,928	8.1

Figure 3-12
Average Vehicle Age by Category

Vehicle Type	Total	2001	2000	1999	1998	1997	Other
PTL Standard	4	1	-	1	2	-	-
Under 1/2-Ton Pickup	3	-	1	1	-	-	1
1/2-Ton to 3/4-Ton Pickup	36	15	3	5	4	2	7
Service Vans	5	1	1	2	-	1	-
Under 10-Yard Dump Truck	15	4	3	2	4	1	1
Over 10-Yard Dump Truck	3	-	-	1	-	1	1
Truck, General	4	1	-	-	2	-	1
Refuse Compactor	2	1	-	-	-	1	-
Tank with Attach.	6	1	2	-	1	-	2
Trailers	19	2	3	2	4	1	7
Misc. Equipment	13	1	-	4	3	-	5
Air Compressors	3	-	-	-	1	-	2
TOTAL	113	27	13	18	21	7	27

Figure 3-13
Equipment Inventory Summary by Class

- Cost of $ + Physical Inspection + Cost of Old Vs. New Lease Vs. Own Vs. Rent
- When Do I Buy? When Money Is Available!
- Do I Downsize?
- Do I Replace My Underutilized Vehicles?
- Can I Rebuild Rather Than Buy? Lease? Rent?
- Can I Combine Functions on Equipment to Better Utilize?
- Do I Use Vehicles Efficiently for Different Tasks Rather than Special Vehicles for Certain Tasks?
- Can I Use Smaller More Fuel-Efficient Vehicles to Do the Same Work I Do?
- Can I Buy Used Vehicles?
- Common Vehicles to Reduce Inventory, Training

Figure 3-14
Replacement Analysis

---

1) Present Cost of Money
2) Present Cost of Vehicle
3) Estimate Next Year's Cost of Old Vehicle Based on Historical Cost Escalation
4) Estimate Cost of New Vehicle
5) Physical Inspection and Diagnosis

	Old Vehicle	New Vehicle
1) Labor Cost		
2) Parts Cost		
3) Est. Mileage Next Year		
4) Fuel Cost		
5) Registration $		
6) Tax $		
7) Insurance $		
8) Principal		
9) Interest		
TOTAL		

Minus Salvage Value _____
Total New _____

TOTAL OLD _____

Figure 3-15
Vehicle Replace Versus Repair Analysis

New Pickup 1999	$13,800.00	New Pickup 2003	$19,500.00
Depreciation 50 Month	$ 276.00	Depreciation 50 Month	$ 390.00
Interest 5%	$ 34.50	Interest 5%	$ 48.75
	$ 310.50/Month		$ 438.75/Month
	$ 3,726.00/Yr		$ 5,265.00/Yr

$1,539 Difference per Year

Operating Cost - 15,000-Mi per Year

$1,300	Maintenance Cost	$ 400
$750	10 mpg vs. 20 mpg	$ 750
$5776	TOTAL	$6,415

It is $639 more cost-effective to keep the old vehicle based on a one-year analysis.

Fully Depreciated Pickup

	No Principle or Interest Cost
	No Depreciation
$2,500	Maintenance Cost
$1,500	Fuel @ 10 mpg
4,000	per Year

2003 Year Cost $6,415 — Fully Depreciated $4,000 = $2,415 Difference

If we kept a fully depreciated pickup truck, we would have a $2,415 cushion to fall on rather than buy the new vehicle.

If we compared the same to a 3-year-old (1999) pickup, we would have a $1,776 cushion.

The question is vehicle availability and company cash position.

Figure 3-16
Sample Old vs. New Comparison

this area in your evaluation because it is a product of an ever-changing economic environment. However, look to the controller to feed you this information.

An evaluation of the pickup truck on Figure 3-16 shows several alternatives. Because the 3-year-old 1999 vehicle is $639 cheaper to operate and maintain, it is more cost-effective to keep it and not buy the 2003 model. The fully depreciated vehicle, even if it has 90,000 mi on it, could be more cost-effective to keep because the $2,415 difference of the one new unit could be

New 2003 Diesel Truck		New 2003 Gas Truck	
Depreciation 84 Mo.	$ 55,000	Depreciation 84 Mo.	$ 50,000
Interest 10%	$ 655	Interest 10%	$ 595
Per Month	$ 292	Per Month	$ 238
Per Year	$ 917	Per Year	$ 833
	$ 11,004		$ 9,996

Difference $1,008 Per Year

$ 600 Yr.   Maintenance Cost $ 1,000 Yr.
$ 2,500 Yr.   6MPG   (1.00 Gallon Fuel 15,000 Miles)   3MPG $ 5,000 Yr.

Total Cost Year 1   Total Cost Year 1
Diesel   Gas
$14,104 $15,996

Net Difference
$1,892 Per Year ($158/Mo)
32 Months to Save $5,000 Initial Cost Difference
84 – 32 = 52 Months @ $158 = $8,216 Savings

Figure 3-17
Gas Versus Diesel

wiped out with one major repair. But if you have ten old units, you could keep them rather than buy new and recover the $2,415 per vehicle because the chance of major repairs is less likely to cancel all of the $24,150 savings in the next year.

Many other factors can skew this analysis. However, for illustration, the point is, would a new unit cost less to own and operate than the old? If so, buy new; if not, keep the old.

The analysis continues to boil down to the cost to maintain a vehicle versus the cost of money to buy a new vehicle.

How about accidents? In most cases, if a vehicle repair costs 50% of its present market value, it is time to look at a no-fix alternative.

The controller can deny the new purchase based on ever-changing company profitability and tax laws. Transportation should initiate the request and communicate with the "bean counter," understanding the applied money logic.

Your need is to relate to ownership and maintenance costs in new, rebuild, and do-nothing alternatives (Figure 3-18) and ask yourself: Is it time to buy new,

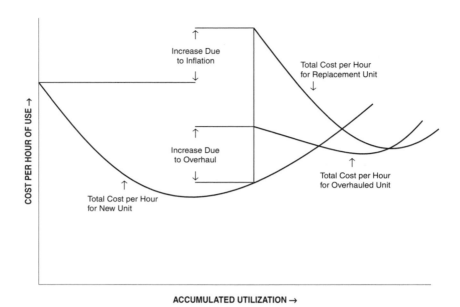

Figure 3-18
Total Ownership and Maintenance Costs
for Overhauled and Replacement Equipment
(Courtesy Byrd, Tallamy, McDonald, and Lewis, Falls Church, VA)

hold capital costs down, rebuild old, or keep old? You can project with reasonable accuracy the vehicle cost pattern of tomorrow based on your life-cycle history accuracy. You need at least a three-year history or a full cycle of historical maintenance and operating data for this task. The controller plugs in cost of money and tax advantages, and you can address the purchase new or keep the present vehicle alternative. Figure 3-14 through Figure 3-18 illustrate this approach. These sample calculations and resultant analysis will bring you closer to the point of making a decision.

Each year, the total annual cost of an in-service vehicle should be compared to the total projected cost of a new vehicle. This will identify comparative cost information for life-cycle cost decision making (Figures 3-19 and 3-20).

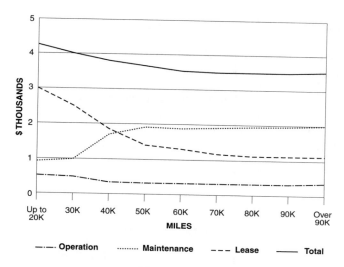

Figure 3-19
Average Cost per Vehicle, Compact Passenger Car
10,000 Mi per Year, Dollars per Thousand Miles

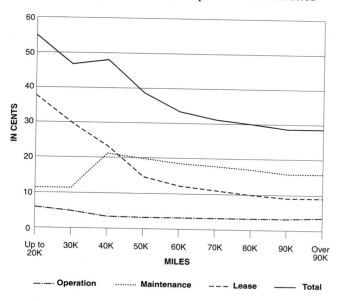

Figure 3-20
Average Cents per Mile, Compact Passenger Car
10,000 Mi per Year, Cents per Thousand Miles

83

Small Vehicle Replacement Analysis

Here is a review of a pickup truck life-cycle analysis using a resale estimate price in an analysis.

3-Year-Old Compact Pickup Truck

Cost	$13,500	New
Principal	$276	(50-Month Term)
Interest 5%	$34.50	(50-Month Term)

Total Payment $310.50 per Month for 50 Months

Operating Costs

15,000 Mi at 20 mpg = 750 Gal of Fuel

750 Gal at $1.30/Gal = $975 per Year

Registration + Misc. Operating Expense = $50 per Year

Total Annual Operating Expense = $1,025 per Year

(Although utilization tends to decrease as vehicles age, a constant annual mileage is used as a base for this analysis.)

Sample Maintenance Cost per Year
(15,000 Mi per Year)

1st Year	2.6¢/mi	$360
2nd Year	5.0¢/mi	$750
3rd Year	8.6¢/mi	$1,300
4th Year	10.0¢/mi	$1,500
5th Year	12.0¢/mi	$1,800
6th Year	13.0¢/mi	$2,000
7th Year	15.0¢/mi	$2,200

Maintenance Costs Tend to Increase as Vehicles Age

Maintenance Costs Increase as Vehicles Mileage Increases

Life-Cycle Cost

Compact Pickup Truck - $13,500

Year	Principal & Interest	+	Maintenance	+	Operating	=	Total
1	$ 3,726	+	$ 400	+	$ 1,025	=	$ 5,151
2	$ 3,726	+	$ 750	+	$ 1,025	=	$ 5,501
3	$ 3,726	+	$ 1,300	+	$ 1,025	=	$ 6,051
4	$ 3,726	+	$ 1,500	+	$ 1,025	=	$ 6,251
5	$ 621	+	$ 1,800	+	$ 1,025	=	$ 3,446
6	$ 0	+	$ 2,000	+	$ 1,025	=	$ 3,025
7	$ 0	+	$ 2,200	+	$ 1,025	=	$ 3,225

Now here are costs for a new pickup truck three years later. Presume the cost of the new truck increases 3% per year.

New Vehicle Increases in Price

3% per Year

3-Year Increase to $19,500 from $13,800

Principal	=	$390.00
Interest 10%	=	48.75
Total Monthly Cost		$438.75

The same maintenance and operating costs are used for this sample.

Year	Principal & Interest	+	Maintenance	+	Operating	=	Total
1	$ 5,265	+	$ 360	+	$ 1,025	=	$ 6,650
2	$ 5,265	+	$ 735	+	$ 1,025	=	$ 7,040
3	$ 5,265	+	$ 1,095	+	$ 1,025	=	$ 7,590
4	$ 5,265	+	$ 1,215	+	$ 1,025	=	$ 7,790
5	$ 877.50	+	$ 1,380	+	$ 1,025	=	$ 3,702
6	$ 0	+	$ 1,560	+	$ 1,025	=	$ 3,025
7	$ 0	+	$ 1,665	+	$ 1,025	=	$ 3,225

Presume a vehicle is reduced in its sales value 30% of the initial cost the first year and 20% of the residual cost each year thereafter. This is a textbook forecast.

The real fact to be considered here depends on the economy—resale values can change.

If new vehicles cost too much, people will not buy them. This will cause the resale price of used vehicles to remain higher because they are in more demand by the buying public.

If new vehicles have a high incentive such as rebates and lower finance rates, then resale values of used vehicles will be reduced.

This is why life-cycle costing is an annual exercise. You must factor in the economy.

Vehicle Net Worth Is Generally Based on a 30% Reduction in Retail Price in Year #1 and 20% per Year Reduction on the Residual Value for Each Year Thereafter

Vehicle Net Worth		Retail Reduction		Residual Value	End of Year
$13,800.00	×	30%	=	$9,660.00	#1
$9,660.00	×	20%	=	$7,728.00	#2
$7,728.00	×	20%	=	$6,182.40	#3
$6,182.40	×	20%	=	$4,945.92	#4
$4,545.92	×	20%	=	$3,956.74	#5
$3,956.74	×	20%	=	$3,165.39	#6
$3,165.39	×	20%	=	$2,532.31	#7

Analysis:

Look at Year 4 replacement and calculate the total cost of the old vehicle for Years 4, 5, 6, and 7 compared to the total cost of the new vehicle for Years 1, 2, 3, and 4.

Life-Cycle Cost Would Be
Cost of Old Versus Cost of New

Present Vehicle		New Vehicle	
Year 4	$6,251	Year 1	$6,650
Year 5	$3,446	Year 2	$7,040
Year 6	$3,025	Year 3	$7,590
Year 7	$3,225	Year 4	$7,790
Total 4 Year Cost	$15,947	Total 4 Year Cost	$29,070

4-Year Resale Value of the Present Vehicle $6,182

New Vehicle Net Cost $22,888

The present vehicle is $6,941 more cost-effective over a 4-year period than the new vehicle mainly because of its reduced principal and interest costs—$1,735 per year average savings.

Analysis of Lower Principal Replacement

Subtract the resale value of the old vehicle from the principal of the new vehicle and analyze the results.

If New Vehicle Costs $19,500

(3% per Year Price Increase, for 3 Years on $13,800)

Resale of Present Vehicle at the End of Year 3 is $6,182

(30% Year 1 + 20% Year 2 + 20% Year 3)

The Net New Vehicle Cost Is $13,318

Principal for 50 Months	$ 266.36
Interest at 5% per Year	$ 28.48
Monthly Cost	$294.84

Year	Principal & Interest	+	Maintenance	+	Operating	=	Total
1	$ 3,586	+	$ 400	+	$1,025	=	$ 5,011
2	$ 3,586	+	$ 750	+	$1,025	=	$ 5,361
3	$ 3,586	+	$ 1,300	+	$1,025	=	$ 5,911
4	$ 3,586	+	$ 1,500	+	$1,025	=	$ 6,101
5	$ 532	+	$ 1,800	+	$1,025	=	$ 3,357
6	$ 0	+	$ 2,000	+	$1,025	=	$ 2,525
7	$ 0	+	$ 2,200	+	$1,025	=	$ 3,225

Life-Cycle Cost Analysis Includes New Vehicle Purchase Price
Minus Present Vehicle Resale Value

Present Vehicle Cost		New Vehicle Adjusted Cost	
Year 4	$6,251	Year 1	$5,011
Year 5	$3,446	Year 2	$5,361
Year 6	$3,025	Year 3	$5,911
Year 7	$3,225	Year 4	$6,101
Total 4		Total 4	
Year Cost	$15,947	Year Cost	$22,384

The present vehicle is $6,437 more cost -effective over Years 4, 5, 6, and 7 than the new vehicle in Years 1, 2, 3, and 4 ($1,484 per year average savings).

Year 1	Deficit	–	$1,240
Year 2	Savings	+	$1,915
Year 3	Savings	+	$2,386
Year 4	Savings	+	$2,876

Another life-cycle cost alternative is to make a decision to repair, rebuild, or replace a vehicle when the maintenance cost of a one-time expense or an annual accumulated expense equals 30% of the residual value of the vehicle. This is the time to inspect the vehicle to determine what is needed to extend its life prior to an unscheduled rebuild cost (Figure 3-21 and 3-22).

	Year 1	Year 2	Year 3	Year 4	Year 5	Year 6	Year 7	Year 8
Principal	$4,440	$4,440	$4,440	$4,440	$740	—	—	—
Interest	$925	$703	$481	$259	$37	—	—	—
Parts/Labor	$360	$735	$1,095	$1,215	$1,380	$1,560	$1,665	$6,200
Fuel	$480	$480	$480	$480	$500	$500	$500	$500
Cents per Mile 15,000	0.414	0.423	0.433	0.426	0.197	0.137	0.144	0.446
Cumulative Average		0.418	0.423	0.423	0.378	0.338	0.310	0.327
Resale	$12,950	$10,360	$8,288	$6,630	$5,304	$4,243	$3,395	$2,716
Percent Maint. Residual	3%	7%	13%	18%	26%	37%	49%	228%
						↑		↑
						Replace Target Evaluation		Must Replace or Repair or Rebuild

Figure 3-21
$18,500 Light Vehicle

	Year 1	Year 2	Year 3	Year 4	Year 5	Year 6	Year 7	Year 8	Year 9	Year 10
Principal	$14,000	$14,000	$14,000	$14,000	$14,000	—	—	—	—	—
Interest	$3,500	$2,800	$2,100	$1,400	$700	—	—	—	—	—
Parts/Labor	$1,455	$2,475	$3,780	$3,690	$3,495	$3,375	$3,800	$4,500	$8,500	$22,425
Fuel	$3,600	$3,600	$3,600	$3,600	$3,600	$3,600	$3,600	$3,600	$3,600	$3,600
Cost per 15,000 mi	$1.50	$1.53	$1.57	$1.51	$1.45	$0.47	$0.49	$0.54	$0.81	$1.74
Resale	$49,000	$39,200	$31,360	$25,088	$20,074	$16,056	$12,845	$10,276	$8,221	$6,577
Percent Maint. Residual	3%	6%	12%	15%	17%	21%	30%	44%	103%	341%
							↑			↑
							Replace Target Evaluation			Must Replace or Repair or Rebuild

Figure 3-22
$70,000 Chassis—Heavy-Duty

Another life-cycle approach is when the accumulated maintenance costs for the vehicle equal its purchase price (Figure 3-23). Now it is time to replace the old with the new.

Life-cycle costs are broken down into annual costs, which are cash dollars for that calendar year. The final evaluation stage is present-value dollars, the year the vehicle was purchased, and each year thereafter converted to the dollar value of the purchase year or, better yet—present value levelized—that year after purchase multiplied by the years since purchase to quickly see which year forecasted is our best value.

In Figure 3-24, the levelized column indicates that if we keep this textbook vehicle 12 years and sell it at the end of the twelfth year, it will have cost us $22,353 per year for 12 years to own and operate. In Figure 3-25, if we kept this vehicle 12 years, sold it in the tenth year and bought new for 2 years, our 12-year levelized cost is $24,018 per year for 12 years.

89

Year	LTD Miles	Cumulative		LTD Maintenance Accumulated Cost	Actual	
		LTD Accumulated CPM			Annual Maintenance Cost	Annual CPM
1	100,000	2.50c		$2,500	$2,500	2.50c
2	200,000	3.50c		$7,000	$4,500	4.50c
3	300,000	4.50c		$13,500	$6,500	6.50c
4	400,000	5.50c		$22,000	$8,500	8.50c
5	500,000	6.30c		$31,500	$9,500	9.50c
		1¢ more than LTD		↑		
		2¢ more than LTD		**Accumulated Maintenance**		
		3¢ more than LTD		**Costs Equal Vehicle's**		
		3.2¢ more than LTD		**Purchase Price**		

Figure 3-23
Annual and Life-to-Date (LTD) Maintenance Cost per Mile (CPM)
(Original Vehicle Cost $31,500)

Year	Lease Payment ($)	Maintenance Cost ($)	Operating Cost ($)	Resale Value, Year End ($)	Annual Cost ($)	Present-Value Annual Cost ($)	Cumulative Present-Value Annual Cost, Levelized ($)
1	24,000	1,000	500		25,500	22,090	25,500
2	24,000	2,000	500		26,500	21,354	25,973
3	24,000	3,000	500		27,500	19,892	26,482
4	24,000	4,000	500		28,500	18,506	26,856
5	24,500	5,000	500		29,500	17,195	27,285
6	24,500	5,500	550		30,050	15,723	27,631
7	24,000	6,000	550		30,550	14,349	27,926
8	600	6,500	550		7,650	3,225	26,241
9	600	7,000	550		8,150	3,085	24,985
10	600	7,500	550		8,650	2,939	24,027
11	600	8,000	550		9,150	2,791	23,283
12	600	8,500	550	8,000	1,650	452	22,353
					233,350	141,601	
					19,446/yr	11,866/yr	

Figure 3-24
Vehicle Cost Analysis Replacement at 12 Years

In a sample 12-year window—in a present-value levelized format in a brief analysis—the 12-year life cycle of this textbook vehicle sold in the end of the twelfth year is our lowest cost per year to own and operate this vehicle.

Summary

Vehicle life-cycle costing depends on the cost of new versus the cost of old. Major factors to evaluate costs are:

Year	Lease Payment ($)	Maintenance Cost ($)	Operating Cost ($)	Resale Value, Year End ($)	Annual Cost ($)	Present-Value Annual Cost ($)	Cumulative Present-Value Annual Cost, Levelized ($)
1	24,000	1,000	500		25,500	22,890	25,500
2	24,000	2,000	500		26,500	21,354	25,973
3	24,000	3,000	500		27,500	19,892	26,482
4	24,000	4,000	500		28,500	18,506	26,856
5	24,000	5,000	500		29,500	17,195	27,285
6	24,000	5,500	550		30,050	15,723	27,631
7	24,000	6,000	550		30,550	14,349	27,926
8	600	6,500	550		7,650	3,225	26,241
9	600	7,000	550		8,150	3,085	24,985
10	600	7,500	550	15,000	–6,350	–2,157	24,127
11	30,000	1,500	525		32,025	9,767	23,591
12	30,000	3,000	525		33,525	9,178	24,018
					273,100	153,007	
					22,758/yr	12,750/yr	

Figure 3-25
Vehicle Cost Anaylsis Replacement at 10 Years

- Principal and interest of new and old

- Operating cost of new and old

- Resale cost of old

- Incentives for new

- Finance applications—tax credits, company structure growing, staying the same, consolidating (i.e., reduction in the number of vehicles needed)

- Strategic plans for tomorrow, tactical plans for today

Vehicle life-cycle replacement is an annual financial calculation of present vehicle cost to own and operate it versus the cost to own and operate a new vehicle. It is also dependent on:

- Company conditions
    Growth
    Status quo
    Consolidation

- Effectiveness of its vehicle maintenance program

- Maintenance staff skills

- Vehicle utilization

- Vehicle availability

- New vehicle incentives and rebates

- New vehicle buy-back opportunities

- Vehicle obsolescence

- New vehicle technology

- Vehicle resale value

- Configuration of vehicle

- New vehicle cost

- Interest rates

- Company image

- Vendor maintenance support

- Company expensing or capitalizing vehicle costs

- Work method changes

Today, engineers are making a better product. Their better designs are reducing your maintenance costs. Improvements such as fuel injection, electronic control units, anti-lock brake systems, improved ignition systems, better miles per gallon, air deflection for highway driving, fan clutches, better emissions, multiplexing, automatic transmissions, air-operated disc brakes, and a host of other continuing developments have focused on lower maintenance costs.

This impacts your life-cycle analysis because lower maintenance costs motivate the fleet users to extend vehicle life.

Of course, image is a factor in sales and marketing, and "perk" vehicles are salary enhancements. Generally you should examine your vehicle needs annually to best evaluate the pertinent factors in vehicle life-cycle costs.

Shop and Building Equipment Life Cycle

A review of the equipment installed in maintenance facilities has the following life-cycle estimates for a useful reference. These estimates are based on hours and years. As far as hours as a reference, I have used 52 weeks × 8 h (1 shift), which is equal to an annual use of 2080 h. Two shifts of operation during 1 calendar year would be 4160 h or 2 years useful life. Each piece of equipment will be defined for the criteria using one shift per day as a reference.

Equipment	Years	Hours
Sweepers	10	20,800
Hydraulic Lifts	20	41,600
Cranes	20	41,600
Chassis Dynamometer/Brake Analyzer	10	20,800
Facility Management System	5	10,400
Wheel Alignment	20	1,600
Bus Washer	15	31,200
Steam Washer	10	20,800
CNG Fueling Equipment	10	20,800
HVAC Equipment	10	20,800
Spray Paint Booths	15	31,200
Emergency Generator	15	31,200
Fire Alarm, CCTV	10	20,800
Alternator Test Stand	15	31,200
Medium-Duty Lathe	20	41,600
Dual-Spindle Brake Lathe	20	41,600
Drill Press/Cutoff Saw/Misc.	20	41,600
Portable Equipment/Presses/Radial	20	41,600
Drill/Battery Room	20	41,600
TUG – Push–Pull Tow Vehicle	10	20,800
Welding Equipment	15	31,200
Diesel Fuel Monitoring System	10	20,800
Fluids: Storage and Waste Inventory	10	20,800
Forklift/Stacker: Certification	10	20,800
Parts: Cleaning and Storage	15	31,200
Electronics Test Equipment	10	20,800

A periodic review of technological improvements of this machinery and equipment would, in fact, impact productivity resulting from the use of this improved equipment. If issues surface that show replacement of the above by productivity increases, a cost analysis should be initiated to show the benefits on the return of the dollars invested in the cost of the new product, which would be the basis for a return on the investment justification to replace the old with the new, even though the old is still usable. Although usable, it is not the most productive and should be replaced by the new.

Facility Equipment Life Expectancies

Equipment/Subsystem	Normally Expected Service Life
Steam Unit Heaters	20 years
Electric Unit Heaters	13 years
Steam Radiators	25 years
Steam Heating Coils	20 years
Sheet Metal Ductwork	30 years
Duct Dampers	20 years
Fans:	
Centrifugal	25 years
Axial	20 years
Propeller	15 years
Rooftop Ventilators	20 years
Pneumatic Controls and Actuators	20 years
Self-Contained Equipment Controls	10 years
Base-Mounted Pumps	20 years
Inline (Pipe-Mounted) Pumps	10 years
Sump Tanks and Sump Pumps	10 years
Condensate Pumps	15 years
Electrical Transformers	30 years
Shell and Tube Heat Exchangers	24 years
Window Air Conditioners	10 years
Standard Fluorescent Lamp Ballasts	20 years
Sanitary Waste Piping	40 years
Heating/Cooling	30 years

Chapter 4

Vehicle Sales

╬

Resale Strategy

As part of fleet management life-cycle economics, the sale of replaced, underused, high-cost accident vehicles is a topic that is essential to each fleet manager's profile. Each class of vehicle—be it light trucks, medium vocational, heavy vocational, trailer, or construction equipment—requires a different strategy that is driven by the market.

However, to sell your vehicle, the vehicle must be desirable. Good color scheme, mileage, exterior conditions, no accident history, and extra accessories all contribute to the perception or desirability of the vehicle.

Buyers want these vehicles and look to save the amortized value of the vehicle.

For example, let us say a car costs $25,000 to purchase. The textbook depreciation schedule is 30% reduction for the first year, making the residual value $17,500 if this vehicle is put up for sale and is within average yearly mileage (15,000–20,000 mi). The attractiveness to the purchasing customer is the $7,500 savings. If a vehicle is put up for sale at 5 years, the $25,000 purchase price is depreciated to $7,168 (30% reduction Year 1 and 20% each year thereafter) and if the mileage is between 50,000 and 75,000 mi, we have an attractive $17,832 perceived savings by the purchasing customer if the vehicle has been cared for and properly maintained.

In the last 10 years, technology has made cars better. In fact, all classes of vehicles have improved. Therefore, keeping a vehicle longer, if it is operated and maintained properly, provides a saleable product with a potential of more than 2000 mi of reliable service. If the vehicle is garaged and serviced properly, more years and more miles are possible, and less maintenance costs are experienced by the initial owner, making it an attractive option to hold onto a vehicle for more than six years and still obtain some good net value for resale. Also, the marketing perception of the stockholders is based on conservatism. Thus, it is perceived that maintaining older vehicles means more potential return on their investment.

Beware of Liability

Liability issues must be considered. When you sell a vehicle, you are liable for its safety to the purchaser and all future purchasers. With the judicial process being favorable to the purchaser, *"caveat emptor"* or "let the buyer beware" is not as strong as it was 20 years ago. In fact, if a vehicle were sold by a utility directly to the customer or through an auctioneer or third party, and several years later an accident occurs that is thought to be caused by a maintenance or design problem, the utility stands an excellent chance of being put into the litigation loop with previous purchasers. The utility could be forced to show due diligence to provide a safe vehicle to be sold. Regardless, after the process, the utility could pay part of the settlement costs only because it has "deep pockets," even if the utility has taken some reasonable precautions, such as having the maintenance records show performing maintenance to the manufacturer's recommendations, repairing design problems in recall and repair campaigns, and having the vehicle pass state inspection annually.

If the vehicle does not pass state inspection, it should be repaired prior to sale and noted so in its records. The records should be kept by the salesperson or company and held for seven years. Should an accident occur, a visual inspection by the customer is sufficient to understand the quality and condition of the vehicle and its perceived reliability and value.

If the vehicle is unsafe or could be perceived as being unsafe due to visual indices of the accident, a secondary approach should be taken and an evaluation of the risk determined. If the vehicle were repaired, the accident not recorded, and then the vehicle sold later, that omission places the liability on the seller. Now we put ourselves in a vulnerable position of incurring the litigation costs of a judicial trial or a pretried settlement that could negate all the revenue generated from our previous sales.

Even if we had a disclaimer that said, "*caveat emptor*, due to normal wear and tear, this vehicle can no longer be safely utilized by Company A employees for its intended purpose," we still remain at risk.

The risk here is why did the utility knowingly sell an unsafe vehicle to a person who might not understand the meaning of such clauses?

Even if we said the left front brake pad should be replaced, which is a simple generally understood statement by the majority of the population, it does not offer liability relief, and the resulting litigation award could jeopardize our fleet economics.

The more complex the vehicle, the higher the risk of accountability for litigation exposure. Win or lose, it costs money to have a professional attorney address these issues. In fact, most utility legal staff cover the initial response to claims. Then, if the claims continue at a higher level, the utility will engage the services of a specialty attorney to decrease the cost of settlement to the utility.

The Alternatives

What are some viable alternatives that reduce the risk of liability? Quite frankly, the best option is to sell the vehicle back to the chassis and mounted equipment supplier. In the purchase of all classes of vehicles, the manufacturers and/or distributors are qualified to inspect reconditioned, pre-owned equipment and assume the risk of resale. The new vehicle specification and boiler-plate terms and conditions would indicate that a proposal for a buyback be stated. Now the seller, knowing the market for aged equipment, can project returns, thus giving the utility fleet manager a number to plug into the life-cycle analysis to identify cost efficiencies.

The bid response would say, "Vehicle A will cost you X amount of dollars for the first 10 years, X amount of dollars for each of the next 10, etc. If replaced at the end of X amount of months and/or X amount of miles, it will be purchased back by the seller for X amount of dollars." Now the utility budget people have a number with which to work. If they remain with that number, this expense is the capital and operating costs that would be associated with that expense and resale income. If the expense goes outside that number, this would be the cost alternative.

The market changes due to the state of the economy and technology. Today, because of successful vehicle electronics, computers on engines and transmissions (e.g., electronic control units, integrated circuiting from engines to transmissions), antilock brakes, mounted body and equipment systems, multiplexing, and so forth, many fleets sold their old equipment in the early and mid-1990s. Thus, a glut of used equipment has decreased resale value during the 1990s.

Each fleet should have a defined life-cycle program that indicates capital and operating dollars that are needed to remain the same vehicle average age by class. What would the capital and operating costs be if they acquired bigger, smaller, older, or younger, including staff and space costs to support the life cycle needed and any outsource funding to cover expansions?

In doing these calculations for hundreds of fleets, resale value of equipment plays an important part in the economic outcome. If a fleet turns over its units, taking advantage of buybacks from its suppliers, the fleet can recover high resale values to offset the higher capital costs to purchase more frequently. Should the fleet choose to increase the average age of its vehicles, the fleet will lower capital costs and part of the capital savings will move to the increased operating support services side of the budget.

When you get into rebuilding to extend a vehicle life cycle, for each increased work order dollar, you will need parts dollars to support component rebuild costs or replacement component costs.

For example, an average gasoline engine might cost $6,000–$8,000, with labor being $2,000–$3,000 and parts being $4,000–$5,000. Transmissions at $3,000 per replacement could cost $1,000 for labor and $2,000 for parts. Large diesels, rebuilt or new, could run $20,000 with labor at $5,000 and parts at $15,000 with core credit.

Preparing the Vehicle

Resale is an important part of the economic cycle in fleet management. It costs money to prepare a vehicle for sale, which reduces the net profit. If the $25,000 vehicle we discussed is projected for a $7,168 return price, we need to prepare it for sale.

Prior to the sale, the vehicle must be cleaned inside and out for $300, transported to a site for $250, secured at the site for $250, and administered in-house or through a vendor for sale for $500, not to mention the state inspection cost of $250 plus the parts and labor to upgrade the vehicle for $600 to pass the inspection. Now the accumulated costs per vehicle as estimated previously at $2,150 subtracted from the $7,168 expected price bring down the net sale price to $5,018. Should an auctioneer sell it for us? The average of 8% would be charged for actual sales price: $7,168 × 8% = $573.44 + $2,150 prep costs, which brings down the revenue to $4,445, or 62% of the projected sales price.

The "rule of thumb" is that you will receive 62% of the projected sales cost, whether outsourced or done in-house. The buyback looks so much more efficient when you break out the costs incurred in this process.

Prepping to Avoid Litigation

In preparing a vehicle for sale, evaluating its individual condition is of primary importance. Photos of the vehicle should be taken, with an assessment of its perceived condition. Similar to its specific preventive maintenance (PM) inspection sheet (which also should be included) with the work generated from the inspection, an overall assessment should be documented.

Resale is here to stay. We cannot delegate the responsibility or the liability. We can have an auctioneer and manufacturer do the work, but ultimately we are responsible, accountable, and liable for the safety of the vehicle. The fleet manager is the person delegated by the fleet to do this efficiently. The "buck" stops there.

Good and Bad Selling Points in Trucks

What makes a used truck saleable?

- Attractive color scheme
- Cleanliness
- Good maintenance condition
- Good matched tires

- Good suspension; level chassis
- Tight fittings without leaks
- Good start-up and smooth engine operation
- Good performance under a load
- Clean body: dents and decay repaired

What hurts the chances of a used vehicle being sold?

- Low horsepower engine
- Non-sleeper cabs
- High mileage
- Poor maintenance
- Excessive special equipment
- Short wheelbase (for highway applications)
- Long wheelbase (for city applications)
- Spoke wheels
- Decals
- Poor paint
- Body damage

A used vehicle cannot appeal to all potential buyers. The majority of buyers are small companies or owner-operators who compete for what is available. When money is expensive to borrow, many fleets hold on to their equipment longer or obtain used equipment to buy time until the economy improves. A well-maintained vehicle, therefore, offers wide appeal if it is equipped to do the job for which a buyer is seeking a vehicle. Many small fleet buyers will buy used equipment to benefit from the depreciation loss rather than incur higher vehicle new costs. This motivates one to buy a well spec'd out used vehicle and to avoid the new depreciation loss.

Disposal of Used Vehicles

Excess vehicles have to be disposed of. The following are some cost-effective alternatives.

Auctions

By strict definition, it would seem that this practice offers a simple answer to used vehicle disposal. After all, the purpose of an auction is to sell a vehicle at the highest possible price. But this certainly is an oversimplification. An auction, for instance, does not guarantee to assemble all the potential buyers. Then, too, prevailing market conditions in the area where the auction takes place may not be as satisfactory to the fleet operator as market conditions that exist in another locale.

The auction is the method on which several major leasing companies rely as their primary vehicle disposal technique. However, auctions permit little imagination or flexibility in the disposal of fleet vehicles. The auction solution is a quick solution to sell many vehicles, and because of the size of a large fleet, many companies are forced to rely on this method of moving huge quantities of used equipment through clogged pipelines. For fleet operators, there is a major deficiency in the auction disposal profits. You must settle for the auctioneer's wholesale price rather than a larger selling price that could be obtained by selling the vehicles in smaller, more individual lots.

Another drawback to the auction method lies in the time and expense required to accomplish a sale. It can take as long as 45 to 60 days to assemble and monitor used vehicles through various auctions throughout the country. Auction fees and expenses are, of course, paid from the fleet budget and are based on a percent (usually 6–8%) of total auction revenue. The more the vehicles sell for, and the more vehicles actually sold, the higher the auctioneer's commission and the greater revenue for the fleet.

Auctions have several sides to their operations. The price of a vehicle is governed by supply and demand. The auctioneer can create demand by "shilling" the crowd. Representatives of the auctioneer buy the first 10 or so vehicles at inflated prices to set the level of price for the following groups of vehicles. Another example is to start the sale with the best conditioned vehicles first, setting the bidding pace at a higher level for the less desirable vehicles to follow.

At the end of the auction, any vehicles that remain, owing to lack of crowd response, are sold in a group to repeat auction customers as a good-will business gesture.

Auctioneers are aware of the market value of vehicles. There is the present retail price, the highest; then the wholesale price, the middle price; and the finance price, the lowest, which is the value the vehicle would have as collateral on a loan.

You should be aware of these benchmarks. The auctioneer will fluctuate the price starting point to reward repeat volume customers by selling vehicles to them at prices somewhere between the finance and wholesale levels. Buyers with occasional business will be given opportunities close to or over the wholesale prices.

Companies that sell large inventories of vehicles at auction should have their personnel present at their auctions. Their physical presence ensures fair dealings, and they can buy back vehicles if they are not satisfied with the auctioneer's price.

Again, working relationships with the auctioneer are a function of the volume of business a company does with the auctioneer. The more business you do, the greater your considerations. Thus, a small fleet and a large fleet figure in this overall picture in different ways. One common point is that your physical presence at the auctioning of your vehicles will increase your chances of ensuring that the auctioneer's dealings favor you.

You should understand the following auction definitions and laws:

1. **Sale in lots**. In an auction sale where goods are sold in lots, each lot is the subject of a separate sale.

2. **When is the sale complete?** A sale by auction is complete when the auctioneer so announces by the fall of the hammer, or saying "sold." When a bid is made while the hammer is falling in acceptance of a prior bid, the auctioneer may, at his discretion, reopen the bidding or declare the goods sold under the bid on which the hammer was falling.

3. **Items sold with, or without, reserve.** Any auction sale is hold with reserve unless the goods are in explicit terms put up without reserve or on an absolute basis. In an auction with reserve, the auctioneer may withdraw the goods at any time until he announces the completion of the sale. In an auction without reserve (absolute), after the auctioneer calls for bids on an article, or lot, that lot cannot be withdrawn unless no bid is made within a reasonable time.

4. **Bidding by or for the seller.** During an auction without reserve, and the auctioneer knowingly receives a bid on the behalf of the seller, or the seller makes or procures such a bid, the buyer may, at his option, avoid the sale or take the goods at the price of the last good-faith bid prior to the completion of the sale.

5. **Minimum Price Auctions**. Each vehicle has a minimum price posted on the vehicle to inform the attendees that this vehicle will not be sold for less than this price.

Auctioneers are paid on a percentage of gross revenue of sales. The higher the gross sales revenue, the higher their fee. The more you sell these vehicles for, the more you net for your sale.

An example would be 6–8% for the auctioneer, which includes advertising, assembling the bidders, mailing, conducting the sale, miscellaneous expenses associated with the sale, and 92–94% of sale net proceeds going to the fleet owner.

Sell Back to Manufacturer

When you purchase the vehicle new, an opportunity for a buyback agreement is made on a percent of the purchase price for sale for each year a vehicle ages or miles the vehicle accumulates. This allows a predictable resale for timely replacement cycles.

Sell to New-Truck Dealerships

There are occasions when a dealer has a particular need for used vehicles. At such times, dealerships do represent a sound disposal outlet. It is also true that this method ensures instant disposal for the fleet. It is a sound policy to request the dealer to bid on the turned-in used truck. This tests that particular market and ascertains the dealer's needs.

However, the drawbacks inherent in exclusive use of this method far outweigh the advantages. The practice of selling back vehicles to the dealership from which they were purchased requires consideration of the initial purchase price. Was the purchase price higher than average? Did you buy parts and service from this dealer during the life of the vehicle? This overpayment is theoretically recouped because the same dealer repurchases the unit at a higher price than competitors would to keep your business. Even though the "overpayment" is theoretically recouped when the same dealer buys back the vehicle after, say, two years, it should be pointed out that during those two years, the excess payment is dormant and probably capitalized. In all probability, rent is paid on the vehicle, too. It is an unnecessary expense, notwithstanding the lost earnings return for the fleet operator.

Sell to Used-Truck Dealers

In certain markets, and under certain conditions, this disposal outlet will pay the most. However, it is absolutely essential that anyone disposing of trucks through used-truck dealers be thoroughly experienced, not only with the specific outlet with which they are working but, in addition, with the "ground rules" and practices of the used market in general. There are good truck dealers, and there are bad truck dealers. The good ones are legitimate dealers. They have been in business for a long time, have developed repeat sources for business, and operate under a code of ethics. The questionable ones, however, cannot be counted on for this high level of integrity. As the problem is, it is often hard to tell the good from the bad. Without a great deal of experience in this market and an inherent "horse-trading sense," the fleet manager will find that this market strategy is extremely dangerous.

Used-truck dealers provide an outlet where experienced lessors and fleet owners can realize excellent return for their vehicles, but the risk for the uninitiated is unacceptably high.

Sell to Fleet Operators' Employees

The policy of offering "terminated" vehicles to company employees offers a couple of unique and attractive advantages. Unquestionably, it creates favorable employee attitudes toward the company and can, indeed, be deemed as a fringe benefit incentive. It also tends to ensure quick disposal of the used vehicles at such time as termination is necessary.

However, such a policy creates the very real danger of increased costs to the fleet operator during the life of the vehicle because of a strong inclination on the part of the employee to "overmaintain" the vehicle. Also, under such a program, the fleet operator probably will not obtain the best possible price for the vehicle. If the employee perceives an excessive price for the vehicle, the employee's workload may be negatively impacted.

Sell Directly to Outside Individuals

It would seem axiomatic that elimination of the intermediate sales level would provide additional profit potential. However, selling at the retail level is not a practical solution for large fleets because it is extremely time-consuming and costly. Also, such vehicles are continually subjected to vandalism, theft, declining markets, and other problems. In addition, it must be recognized that individual retail buyers tend to shop carefully and purchase only the "sharp" units. This leaves a fleet operator stuck with the balance. On the other hand, an intelligent "wholesale-minded" truck manager knows how to set up package deals where the buyers accept the marginal with the good. The retail sale option is another instance that demands that a lessor or fleet owner be thoroughly experienced to realize acceptable profits when disposing of large quantities of used vehicles.

Wholesalers

1. Agree on a sale price.

2. Define sale-price parameters.

3. Do they provide for pickup, reconditioning, and make ready, and if so, at what cost?

4. Are they financially stable?

Used-Car Dealers

They are generally subject to the same qualifications as wholesalers but are considered a better resale method because there is no middleman and they sell directly to the public. Your major concern here is financial stability and liability absorption.

Used-Car Sale or Purchase Disposal Service

This is an organization that arranges to sell used cars for its clients, handling all aspects from pickup to sale. This is a convenient method, but the company must demonstrate superior revenue generation performance to offset the fee.

Consignment

When you place a vehicle on consignment, you agree to be paid when the vehicle is sold. This can be troublesome, because:

1. There usually is a fee—either a flat amount, percentage of the sale price, or a cost-plus arrangement (i.e., consignee can keep anything over your set price).

2. The consignee may invest in repair without your knowledge, the cost of which may be borne by you.

3. Sale may be delayed because the consignee is holding out for a higher price (and greater commission).

4. Results must offset the consignee's fees.

Inspection/Condition Reports (Vehicle Assessment Report)

Such reports (Figure 4-1 and Figure 4-2) should be an established procedure in every fleet department. These should be done at least once a year and, most importantly, carry the signature of field management personnel. Here are some additional points:

• Immediately before the vehicle is replaced, require a condition report which should be available for later comparison with one provided by the seller, to verify your condition checklist.

• As quickly as possible, bring any discrepancies to the attention of the seller and resolve them.

• Use a camera—digital, Polaroid, or otherwise—to support vehicle identity and condition.

Independent Condition Reports

If your differences with the seller on condition persist over a period of time, then it may be necessary to initiate a special, randomly applied inspection program using an independent agency.

VEHICLE NO.:		VEHICLE MAKE:			MODEL:		YEAR:	
Rating Legend:	5 = Excellent	4 = Very Good		3 = Good		2 = Average		1 = Poor
				Inspection Results				
Description	Awarded Rate	Multiplier		Actual Rate	Max. Rate		Factor - Score	
Section 01 Body/Interior		TOTAL MAX. POINTS 15						10.8
Rust	3	× 20	=	60	100			
Condition	3	× 20	=	60	100			
Accident Damage	4	× 40	=	160	200			
Glass	5	× 10	=	50	50			
General Interior Condition	5	× 10	=	30	50	360	72%	
			Total	360	500	500		
Section 02 Tires		TOTAL MAX. POINTS 15						9.0
Tread Wear	3	× 60	=	180	300			
Sidewall Cracking	3	× 40	=	120	200	300	60%	
			Total	300	500	500		
Section 03 Driveability		TOTAL MAX. POINTS 15						9.0
Road Test	3	× 100	=	300	500	300	60%	
			Total	300	500	500		
Section 04 Brake System		TOTAL MAX. POINTS 15						15.0
Stop Ability	5	× 60	=	300	300			
Emergency	5	× 40	=	200	200	500	100%	
			Total	500	500	500		
Section 05 Steering/Suspension		TOTAL MAX. POINTS 10						8.0
Loose	4	× 20	=	80	100			
Vibration	4	× 20	=	80	100			
Pulling	4	× 20	=	80	100			
Standing Appearance (level/square to road)	4	× 40	=	160	200	400	80%	
			Total	400	500	500		
Section 06 Electrical		TOTAL MAX. POINTS 10						10.0
Once Over Safety Items	5	× 100	=	500	500	500	100%	
			Total	500	500	500		
Section 07 Driveline		TOTAL MAX. POINTS 10						10.0
Noise	5	× 20	=	100	100			
Vibration	5	× 20	=	100	100			
Leaks	5	× 20	=	100	100			
Shifting	5	× 40	=	200	200	500	100%	
			Total	500	500	500		
Section 08 Engine		TOTAL MAX. POINTS 10						8.0
Leaks	4	× 30	=	120	150			
Rough Running	4	× 40	=	160	200			
Noise	4	× 30	=	120	150	400	80%	
			Total	400	500	500		
		TOTAL RATE						
TOTAL 92		AWARDED		3260		TOTAL SCORE	79.80	

Total Score Legend:
3200–4000 Excellent
2400–3200 Very Good
1600–2400 Good
800–1600 Average
0–800 Poor

Awarded Score Legend:
88–110 Excellent
66–88 Very Good
44–66 Good
22–44 Average
0–22 Poor

Figure 4-1
Vehicle Assessment Report
Light-Duty Vehicle

VEHICLE NO:		VEHICLE MAKE:		MODEL:		YEAR:		
Rating Legend:	5 = Excellent	4 = Very Good		3 = Good	2 = Average	1 = Poor		
	Awarded			Actual	Max.			
Description	Rate	Multiplier		Rate	Rate		Factor-Score	
Section 01 Chassis		TOTAL MAX. POINTS = 15						10.8
Rust and Corrosion	3	×	20 =	60	100			
Condition	3	×	20 =	60	100			
Accident Damage	4	×	40 =	160	200			
Glass	5	×	10 =	50	50			
Interior	3	×	10 =	30	50	360	72%	
			Total	360	500	500		
Section 02 Tires		TOTAL MAX. POINTS = 15						9.0
Tread Wear	3	×	60 =	180	300			
Sidewall Condition	3	×	40 =	120	200	300	60%	
			Total	300	500	500		
Section 03 Body Mounted Equipmt.		TOTAL MAX. POINTS = 15						9.0
Dump Body Exterior	3	×	20 =	60	100			
Dump Body Interior	3	×	20 =	60	100			
Tail Gate	3	×	20 =	60	100			
Extension Boards	3	×	20 =	60	100			
Lift Cylinder	3	×	20 =	60	100	300	60%	
			Total	300	500	500		
Section 04 Brake System		TOTAL MAX. POINTS = 15						12.0
Service Brakes	4	×	60 =	240	300			
Emergency Brakes	4	×	40 =	160	200	400	80%	
			Total	400	500	500		
Section 05 Steering/Suspension		TOTAL MAX. POINTS = 15						12.0
Looseness	4	×	20 =	80	100			
Vibration	4	×	20 =	80	100			
Pulling	4	×	20 =	80	100			
Parallel to Ground	4	×	40 =	160	200	400	80%	
			Total	400	500	500		
Section 06 Engine and Driveline		TOTAL MAX. POINTS = 25						16.0
Leaks	2	×	20 =	40	100			
Vibration	2	×	20 =	40	100			
Noise	3	×	20 =	60	100			
Shifting	4	×	20 =	80	100			
Rough Running	5	×	20 =	100	100	320	64%	
			Total	320	500	500		
		TOTAL RATE						
	TOTAL 79	AWARDED		2080	TOTAL SCORE		68.8	

Total Score Legend:
2400–3000 Excellent
1800–2400 Very Good
1200–1800 Good
600–1200 Average
0–600 Poor

Awarded Score Legend:
92–115 Excellent
69–92 Very Good
46–69 Good
23–46 Average
0–23 Poor

Figure 4-2
Vehicle Assessment Report
Heavy-Duty Vehicle

- Such agencies include insurance adjusters and national account service companies.

- Arrange for condition reports to be completed on randomly selected vehicles (or in specific areas if it is a localized problem). Explain your intent to the selected agency.

- Compare reports. Then, using 'clean' used-car values, calculate as shown in Figure 4-3.

How Vehicle Condition Determines Sales Method and Results			
Independent Condition Report		**Seller's Condition Report**	
Value (clean)	$ 6,500	Value (clean)	$ 6,500
Less Estimated Repair	$ 325	Less Estimated Repair	$ 765
Estimated Sale	$ 6,175	Estimated Sale	$ 5,735
Actual Sale	$ 6,000	Actual Net Resale Amount	$ 6,000
Projected Loss	$ 175	Projected Gain	$ 265

Figure 4-3
Used-Vehicle Condition Reporting

Setting Prices

This is the most difficult part of used-car sales, and formulas for establishing prices abound. However, an old bromide applies: sell "as-is," because this is the dealer's preference.

If a dealer handles repairs—which delays resale—then he expects to be compensated, and this is the important ingredient you must factor into your pricing formula when selling "as-is." A good rule of thumb is to lower the price by 30% of the estimated repair. Here's an example:

Clean Value	$ 6,500
Estimated Repair	325
	$ 6,175
Repair Factor	98
Estimated Sale Price (as-is)	$ 6,077

Sales Expenses

Regardless of whom purchases your vehicles, there will be expenses involved, which will vary by method of sale. Such expenses should be regularly reviewed.

These may include:

• Pickup fees
• Cleanup
• Sale fees
• Repairs

Sales Details

When fleet vehicles are sold, you should receive copies of the following reports:

• Final adjustment report
• Condition report
• Depreciation analysis leased car sales
• Billing (leased vehicles)

It is unfortunate that most lessors and many fleet owners are almost totally committed to using one or another of these practices as their sole method of disposal. There is no single superior way that should be used as an automatic policy. The most intelligent policy is a flexible one. Vehicle sales depend on the economy.

When to Sell

The proper time to sell a piece of equipment is before the cost of owning and operating an old vehicle is more expensive than owning and operating a new vehicle.

Many considerations enter into the decision on whether to sell or hold on to a vehicle. New vehicles need fewer repairs than older vehicles. Your cost history will show when replacement is cost-effective. With interest on money at 14 to 21%, it takes longer for it to be cost-effective to buy a new vehicle than to repair and continue to operate an old one. Low interest rates on vehicles (4–6%) encourage shorter life-cycle replacement. However, resale value plays a substantial part in reducing the amount of money that must be borrowed to purchase a new vehicle. When it is time to sell, cosmetic upgrade work brings out the best value the present market will allow. Timing lies at the heart of all vehicle terminations—knowing when the vehicle has become a liability.

It has been shown that if new vehicles are higher in cost, used vehicle prices increase. Resales are highest when new models are out for a while. It is a function of supply and demand (see Figure 4-4).

There are various pricing books available for your reference. Figure 4-5 is a summary of a few of the sources. It is important to choose a source with which you are comfortable.

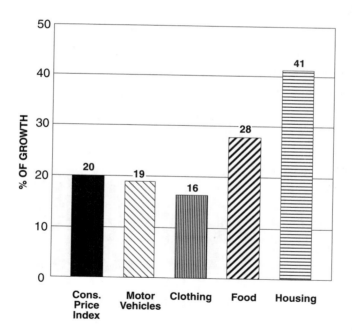

Figure 4-4
Changes in Prices Since 1990–1999
(Source: U.S. Department of Commerce)

Year	Mfr.	Model	Avg. Cost New	Avg. Mileage	AMR Clean/Zone 1	NADA Wholesale (Blue Book)			NADA Official Used Car Guide			Projected	
						Clean	Avg.	Rough	Trade-In	Loan	Retail	Black Book Used	Black Book Projected
1999	Chevy	Lumina	$21,334	33,843	$16,175	$16,400	$16,075	$14,800	$16,700	$16,075	$17,750	$16,735	$15,700
	Dodge	Acclaim	$18,870	28,506	$14,907	$15,300	$14,575	$13,425	$14,925	$14,450	$17,875	$14,240	$13,875
	Dodge	Charger	$21,922	36,149	$16,165	$16,100	$15,250	$14,000	$16,500	$15,550	$17,100	$16,850	$15,075
	Ford	LTD	$24,325	53,441	$18,339	$18,550	$17,450	$15,925	$18,575	$17,725	$19,900	N/A	$17,700
	Ford	Taurus GL	$23,211	47,918	$16,679	$16,900	$16,100	$14,750	$17,050	$16,350	$18,250	$17,775	$16,900
	Merc	Gd Marquis LS	$23,084	41,574	$20,064	$19,850	$18,875	$17,275	$19,700	$18,700	$21,200	N/A	$17,700
	Merc	Sable LS	$25,199	33,338	$17,999	$18,200	$17,300	$15,875	$18,900	$18,150	$20,025	$18,515	$17,800
	Olds	Delta	$25,458	27,330	$18,913	$19,425	$18,525	$17,050	$19,825	$18,950	$21,150	$20,185	$17,600
	Ponti	Bonneville LE	$25,676	35,489	$19,157	$19,175	$18,200	$16,650	$19,700	$18,800	$21,100	$20,240	$17,050
AVERAGES			$23,564	37,510	$17,600	$17,767	$16,928	$15,528	$17,986	$17,194	$19,150	$16,060	$16,489

VALUE Determined on Base + Options +/– Mileage Allowance

Issues Used:
AMR (Clean..Zone 1): 06-14-01
NADA Wholesale (Blue): 06-03-01
NADA OFFICIAL USED CAR GUIDE (Yellow): 07-01
BLACK BOOK USED CAR MARKET: 04-16-01
BLACK BOOK PROJECTED: 12-15-01

Figure 4-5
Resale Value Comparison, 7/01

To sell vehicles to your employees at the highest prices delivers a poor message to them. This is not good business. To sell to the non-employee at the highest prices is good business for your company, because supply and demand of equipment will determine if your sales are successful.

The pricing books vary in prices based on subscriber. One book would be low for auto dealers, another lower for banks, and another higher for used car dealers. You should choose the best level for your use. Figure 4-5 illustrates these differences.

Each book would have several prices per unit. For example, one book has three levels:

	1995 Vehicle	
Retail Value	Wholesale	Finance (Loan)
$4,000	$3,500	$3,000

Another would have categories of:

Clean	Good	Fair	Rough

All have high and low mileage tables to add to or subtract from base costs. In addition, accessories available add to or reduce base costs, such as:

Stereo tape	+$300
Aluminum wheels	+$200
Air conditioning	+$500

Also, different geographic parts of the country have different price levels based on the cost of living variances.

Keeping track of vehicle sales allows for review of strategies and continuation of the best sales method. A company sales program bringing a higher average price per vehicle through its efforts should be compared to the fleet leasing company sales efforts. The different methods must be evaluated so that the best alternative can be pursued.

A review of driver write-ups (Figure 4-7) located in the 90-day dispatch file will show any trends to consider. A policy and procedure should be written and circulated per the following sample (Figure 4-6).

DVIR
Driver Vehicle Inspection Report

29.00   Policy
DVIR-Driver Vehicle Inspection Report

29.01   Procedure
All vehicles will be pretripped prior to movement for damage and federal motor carrier violations. Recorded on previous driver inspection report 396.13C.

At the end of each day's work by the authorized driver, an attached driver vehicle inspection report will be filled out according to 396.11.

    1 cc Vehicle—to dispatch
    1 cc Shop—to dispatcher
    1 cc File—kept at shop

If no repair is needed and noted when next authorized driver reports and a vehicle is assigned, that original report of the vehicle will be given to that driver for review and put into tractor/truck.

If a mechanic repaired the vehicle based on a DVIR, the original and a copy of the work/repair order will be sent to Headquarters Safety after the driver reviews, signs, and puts a copy in the truck/tractor.

John E. Dolce
Fleet Manager
July 1, 2000

Figure 4-6
Driver Vehicle Inspection Report Policy

```
┌─────────────────────────────────────────────────────────────────────┐
│                    Department of Public Woks                          │
│                    Division of Fleet Management                       │
│  Pre- and Post-Trip: Vehicle/Equipment Inspection Form Date: _____ │
│  Department: _____        Driver: _____ │
│  Plate No. _____   Vehicle/Equip. No._____  Mileage/Hour Meter_____ │
│  Oil _____     Exhaust System _____  Other _____ │
│  Transmission_____   Steering_____   Tires-Rims-Wheels _____ │
│  Lights-Wipers _____  Coolant _____   Coupling Devices_____ │
│  Electrical _____  Emergency Equip._____  Brakes _____ │
│  Heater-AC _____   Damage_____   Horn-Mirrors _____ │
│  Comments_____ │
│  _____  │
│  _____  │
│                                                                       │
│  Keep form on file for 90 days; mechanic to sign when work is complete; fleet liaison to sign and │
│  verify work                                                          │
│  Mechanic: _____     Liaison: _____  │
│  Date: _____     Date: _____  │
│  1 cc Dispatch—Vehicle                                                │
│  1 cc Shop File                                                       │
│  1 cc Vehicle File in Dispatch Office                                 │
└─────────────────────────────────────────────────────────────────────┘
```

Figure 4-7
Vehicle/Equipment Inspection Form

Vehicle Inspection

Each vehicle to be sold should be driven and inspected as to condition for reference. Figures 4-8 and 4-9 illustrate a detailed inspection to determine vehicle condition for setting sales price.

A logical inspection process will ensure integrity and objectivity for reference. It will identify vehicles to be made ready and the need for repairs prior to sales to maximize recovery costs.

Reference here is the *AMR Price Book*, "Automotive Market Report."

Vehicle Disposal Liability

Selling a vehicle incurs a liability for its disposal. A vehicle should be safe and have no defects. It should pass the domiciled state inspection prior to sale. Should it have a known defect and an injury or property damage is sustained by the buyer, the seller may be fully or partially liable.

```
┌─────────────────────────────────────────────────────────────────────┐
│  Vehicle Number: _____        Assigned To: _____ │
│  Location: _____          Current Mileage: _____ │
│                                                                        │
│  Year: _____  Make: _____  Model Description: _____ │
│  Color: _____  Exterior: _____  Interior: _____ │
│  Serial Number: _____                   │
│  In-Service Date: _____        Original Cost: _____ │
│  Condition Rating: _____         AMR Issue Date: _____ │
├─────────────────────────────────────────────────────────────────────┤
│  Base Vehicle Includes: AT, PS, AC, AM/FM Stereo, _____ Cy1, _____ │
│  Base Vehicle: ......................................................... $ _____ │
│  Mileage Deduct: Chart_____ ......................................... > _____ │
│  Tape/Cassette: ...................................................... _____ │
│  Tilt Wheel: ......................................................... _____ │
│  Power Windows: ...................................................... _____ │
│  Power Seats: ........................................................ _____ │
│  Cruise Control: ..................................................... _____ │
│  Power Locks: ........................................................ _____ │
│  Other: .............................................................. _____ │
│  Other: .............................................................. _____ │
│  TOTAL COST OF VEHICLE: .............................................. $ _____ │
│  Comments: _____            │
│  Prepared by: _____        Date: _____           │
└─────────────────────────────────────────────────────────────────────┘
```

Figure 4-8
Automotive Market Report

You should inspect the vehicle and state the condition of the vehicle to the best of your ability. If you make no statement, you imply a full warranty and are totally responsible.

Here are two examples of statements for warranty purposes.

The following statement is suggested to be used if a vehicle is in usable shape:

> "This vehicle has been subject to ordinary wear and tear. By retiring this vehicle to you, Company A makes no expressed or implied warranty of merchantability or fitness for a particular purpose."

The following statement is being used if the vehicle is knowingly unsafe:

> "Due to normal wear and tear, this vehicle can no longer be safely utilized by Company A employees for its intended purposes."

```
                        Old Veh. or
     Road Test_____ New Veh._____ Demo _____ Assigned Veh#_____
     Driver_____ Loc._____
     Make_____ Year_____ Model_____
     Mileage_____ Lic No. _____ Other ID_____
     (1)   Exterior: _____
           _____
           _____
     (2)   Engine:  _____
           _____
           _____
     (3)   Transmission: _____
           _____
           _____
     (4)   Steering: _____
           _____
           _____
     (5)   Suspension: _____
           _____
           _____
```

Figure 4-9
Vehicle Evaluation by Driver

Stating that a vehicle is unsafe reduces its sales price. You must accept this reduction of price or open your liability.

Saying "...as is where is..." does not reduce your liability. It is too general. Should a claim be brought against the company, it will cost the company money to defend itself. This must be evaluated in the choice for the sales strategy process.

Spec'ing for Resale

Don't let sticker price determine the best value. There are numerous items to spec that can actually bring more to the blue book price than what you originally paid.

Here is an example of some items to consider:

- **Engines**—"The higher the horsepower, the better. If you get an engine that can be down-/upgraded, be sure your transmission and exhaust can handle an upgrade."

- **Transmissions**—"Thirteen and 18 speeds with overdrive bring a premium at trade-in time." Example: 13-speed versus 10-speed transmission (cost $1,300; resale $2,500) net resale value of $1,200.

- **Engine Brake**—"This option will more than pay you back at trade-in time." Example: (cost $1,500; resale $2,500) net resale value of $1,000.

- **Air-Ride Suspension**—"Many over-the-road trucks have this feature. Without it, your truck will be harder to sell."

- **Aluminum Wheels**—"Not only do they save weight, they're stylish and are a 'must have' for owner operators. They will usually pay for themselves at trade-in time."

- **Raised-Roof Sleeper with Double Bunk**—"You might not think a double bunk is a big issue if you have solo drivers, but many wholesalers simply will not buy a sleeper without a double-bunk setup."

- **Aero Package**—"If you buy an aerodynamic truck, make sure you spec the full aero package. If fairings, for example, have to be added by a dealer at the wholesale level, it's significantly more expensive than adding it as a factory option. The net result would be a big deduction on resale value."

- **Airglide Fifth Wheel**—"This doesn't cost that much as a factory option but has a nice payback." Example: (cost $620; resale $1,300) net resale value of $680.

- Other items that offer good payback are dual exhaust, dual fuel tanks, and stainless steel air cleaners.

Chapter 5

Economics of Vehicle Replacement

Economics of Vehicle Replacement

A piece of equipment has a number of "lives." It has a service life, which is the amount of time the vehicle is capable of rendering service. This life may be nearly infinite if the unit receives adequate maintenance and if worn-out components are dutifully replaced.

A unit also has a technological life, which represents the relative productivity decline of the unit when compared with newer models on the market. For example, a 10-year-old unit may be only 50% as productive as comparable current models. Although the older vehicles are still capable of rendering service (given an unlimited maintenance budget), they are incapable of matching modern performance requirements. Thus, technological life is quite different from service life.

Most of all, economic life is of critical importance to equipment managers. It relates to the total stream of costs associated with the unit over a period of time. Therefore, it impacts both capital and operating budgets. The economic life of a unit refers to the length of time the average total vehicle cost is at a minimum. Total unit expense encompasses all costs associated with the ownership of the vehicle and includes:

1. Depreciation
2. Operations
3. Maintenance
4. Downtime
5. Obsolescence

6. Operator training
7. Costs of carrying parts in inventory for the vehicle (storage, insurance, etc., although not the actual parts costs)
8. Interest
9. Alternative capital value
10. Inflation

During the service life of a truck, cost elements fluctuate—some increase, some remain level, and a few even decrease. Together, they represent the total cost of the unit, and this cost stream is the relevant factor in determining the economic life of the vehicle.

Not Replacing Vehicles Can Be Expensive

Whether money for new vehicles is available is, of course, an important consideration. However, if purchase money is not budgeted annually, the lack of money will cause the age of a fleet to increase, and its maintenance costs will increase. If company administration cuts funds for purchasing new vehicles, managers can offer some powerful arguments for a reconsideration. More staff will be needed to provide for the increased maintenance needs of the vehicles. More parts will be needed. Which is less—the cost to buy and maintain the vehicles at the average age or the cost of spending money on increased staff and parts?

These arguments are enhanced if records are available to project next year's costs. Compare the increased maintenance costs with new-vehicle costs, assuming that the fleet is the proper size and is fully utilized, and therefore will not require unusually large numbers of purchases. Any excess vehicles could be sold to generate additional new-vehicle purchase funds. Also, new vehicles can be introduced into a so-called fully utilized fleet so that two new vehicles will replace three old ones, further lowering overall fleet costs.

If these ideas can be effectively communicated to top management, the question of vehicle purchase money can be readdressed. The key factor, of course, is credible documentation, based on cost records to which sound management principles are applied.

Concentration on only one or another cost factor, without considering the total, will most likely lead the equipment manager unwittingly to choose replacement equipment that produces lower costs in one or more elements but higher overall costs. This is a situation often forced on equipment managers by peculiarities of the budgeting process. For example, sometimes it is more difficult to obtain capital outlay dollars for vehicle replacement than operating dollars for maintenance. In reality, the amount of capital outlay may be far outstripped by the rapidly escalating costs of maintenance and downtime, not to mention costs that result from holding equipment beyond the economically optimum replacement point. Again, it shows that economic decisions normally cannot be made on the basis of a single cost element. Rather, they require that the equipment manager and budget and finance officer sit down and view the total cost picture.

One means of dealing with this problem is to come to budget meetings, armed with good data and analytical support for replacement proposals. Certain business considerations may dictate deviations from an economically optimum replacement program, but such decisions still should be made in light of objective analysis. An example is the business expanding for more work, staying the same size, or consolidating because of less work. You do not want to replace unused equipment if there is no work. Also, if growing, you would keep the old unit to rebuild it to handle the increase in workload. This is preferable to the situation of many equipment managers who promote replacement plans using intuition rather than hard facts. Intuition may lead to proper conclusions, but intuition is easily dismissed during hard-nosed budget reviews.

Determination of Economic Lifetime

Assume your goal is to minimize the total costs associated with ownership of a particular piece of equipment. How do you determine the economic life for such a unit? Should it be replaced when costs start to rise, or when average yearly costs are exceeded?

The answer, of course, is that new vehicle costs must be compared with those of the current vehicle (per mile or per hour or cumulative dollar costs per year). Today, higher fuel economy alone could justify purchasing a new vehicle instead of keeping less efficient, older units, even if the older units are well maintained.

Life-Cycle Costing

Replacing a vehicle is an annual economic decision. The key factors in evaluating an efficient replacement program are:

- Company growth, reduction of work, *status quo*
- Old and new vehicle principal and interest (depreciation)
- Old and new vehicle maintenance cost
- Old vehicle resale value
- Old and new vehicle operating cost
- New vehicle manufacture incentives

When the principal and interest of the old vehicle decrease, usually the maintenance and operating costs increase (Figure 5-1). The increase in maintenance and operating costs is usually less than the decrease in principal and interest. The annual resale value is the tie breaker, with fleet incentives made available by the manufacturer each year. The manufacturer may give incentives on the quantity of the order.

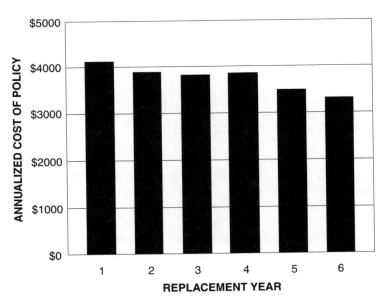

Figure 5-1
Annualized Cost of Replacement Policy

If the resale value of the old vehicle is high enough coupled with the incentive for the new vehicle, it will reduce the principal and interest for the new vehicle. Then if the principal, interest, maintenance, and operating costs of the old vehicle become higher than the principal, interest, maintenance, and operating costs of the new vehicle, you should replace the old vehicle.

If resale values are low with low manufacturer incentives, then this would support keeping the old vehicle and taking advantage of minimal principal and interest costs after the vehicle is depreciated. You will incur a higher maintenance and operating cost by extending the replacement cycle of the vehicle. You should determine if the increased maintenance and operating costs exceed the reduced principal and interest cost savings.

New-vehicle price decreases plus new-vehicle incentives and low vehicle resale prices move fleet managers in the direction of purchasing new vehicles. This could be a result of a downturn in our economy, motivating a change in replacement policy for one year.

There is a need for an annual review for cars and light vehicles because of economic climates impacting company size, resale values, manufacturing incentives, and principal and interest costs. Most large vehicles do not have high resale values or manufacturer incentives to impact economics. If they do because of economic and special applications, you should include these vehicles in an annual analysis to take advantage of any beneficial economies due to these special considerations. Otherwise, you should keep large vehicles until the cost of the old vehicle is more to own and operate than the cost of the new vehicle.

Replacement Policies Within Budget Constraints

Up to this point, it has been implicit that given favorable cost projections, analytical methods, and replacement criteria, sufficient funds would be available for purchasing new trucks when economically feasible. Although some may operate with few budget constraints, the vast majority of equipment managers must live with fund limitations that do not always permit them to replace vehicles at the economically optimum time. Realities of the budget process, legal limitations on borrowing, and administrative policies often leave equipment managers able to replace only a portion of those units due for replacement, if any at all.

Priorities must be set to determine which vehicles to replace with the available funds. The following method is one possible approach.

If a unit is due for replacement, you should project the total cost of the current unit for the following year and compare that cost to the proposed replacement price. The price difference is the basis for not holding the current vehicle beyond its economic point of replacement.

This priority-ranking approach can be developed for an entire fleet by class of vehicle, regardless of departmental assignment. If a given amount is appropriated for replacement of equipment, a *replacement priority rating list* could be constructed to serve as a guide for replacement decisions. Priority ranking is intended to serve as a guide or a tool for decision making. It should in no way be construed as a substitute for the equipment manager's decision-making process.

Many variables lead managers to decide to purchase new vehicles. Limitations on the size of the maintenance facilities may encourage a more vigorous replacement policy in the short run because an older fleet will almost certainly require a greater maintenance commitment. A desire to improve the environmental impact of the fleet may suggest building a newer fleet having cleaner emissions. It may also require outfitting portions of the fleet with special fuel systems or powerplants to reduce pollution. A management desire for a "shiny new fleet image" may dictate shorter replacement cycles. Bureaucratic notions, expectations, and rivalries also affect replacement decisions. Of course, in the long run, economic conditions within the boundaries of an agency influence the manager's decisions, and shortages of revenue and constrained borrowing capacity set the limits on any and all replacement decisions. These considerations, and others, influence decision making and dictate which units should be traded and when they will be replaced.

Vehicle Alternatives

Annually, we should determine for executive management what our cost would be if we remained the same age, grew older, or became younger. We would use capital and operating dollars as a base, noting staff and space adjustments. The older we become, the more repair work is necessary, which means more parts and more labor, which are operating dollars.

Capital dollars determine our life-cycle standard. To become younger, we need more capital dollars than usual and we can give back some operating dollars. The older we become, the less capital dollars we need; however, some of that savings must add to operating dollars to pay for more labor (in-house or outsourced) and more parts.

Part Quality—Life Cycle

1. New fleet:
   - Premature failures
   - Adjust maximum and minimum levels
   - Adjust economic order quantity
   - Warranty

2. Old fleet:
   - Increase maximum and minimum levels
   - Increase line items
   - Decide on rebuilding
       Kits
       Cores
       Inside versus outside
   - Warranty

Supplies
    Uniforms
    Nuts, bolts
    Cleaners, Material Safety Data Sheets (MSDS)
    Rags
    Track levels and cost
    Spread over fleet
    New versus old fleet usage

$15/h mechanic salary, including benefits

Fuel $1.30/gal, 5 mpg, idle 1.5 gal/h

Miles per year 30,000 + 40% idle time
$1.30/gal @ 5 mpg = 6,000 gal × $1.30/gal = $7,800 + 6,000 gal × 40% idle = 2,400 gal × $1.30/gal = $3,120; total = $10,920/yr fuel with 30,000 mi and 40% idle time annually.

Registration annually:
$650 + $10,920 fuel = $11,570

Principle of $105,000 @ 120-mo depreciation = $875/mo

Interest @ 5% = $105,000 Year #1 = $5,250, $94,500 × 5% = $4,725 Year #2

Assumption:
One-half maintenance dollars are parts
One-half maintenance dollars are labor

Example:
Maintenance cost Year #1 = $2,500 = $1,250 parts and $1,250 labor

Mechanic hours = $1,250 ÷ $15 = 83.3 h direct labor

Resale value = 30% reduction Year #1 and 20% per year for each year of its residual value

Example:
$105,000 chassis of mixer × 30% = $73,500 Year #1 and $73,500 × 20% = $58,800 Year #2

As we rebuild and perform major repairs, we will find a ratio of more dollars for parts per our shop labor rate. It is important to know this to forecast properly. Start with a $2 part cost for each $1 of shop labor incurred.

The following textbook example considers 200 mixer trucks as an average of 4 years of age, and compares capital and operating costs to go to a 6-year average age (older) and 2-year average age (younger). This is expressed simply in average operating cost per vehicle to point us toward savings on additional costs.

Vehicle Replacement Strategies and Matching Support Systems

	Year 1	Year 2	Year 3	Year 4	Year 5	Year 6
Principal	$10,500	$10,500	$10,500	$10,500	$10,500	$10,500
Interest 5%	5,250	4,725	4,200	3,675	3,150	2,625
Parts and Labor	2,500	2,000	2,500	3,000	3,500	3,500
Fuel Cost , 30,000 mi	$11,570	$11,570	$11,570	$11,570	$11,570	$11,570
30,000-mi Cost	0.99	0.96	0.96	0.96	0.96	0.94
Resale	73,500	58,800	47,040	37,632	30,000	30,000
Direct Labor Hours	83	67	83	100	117	117

	Year 7	Year 8	Year 9	Year 10	Year 11	Year 12
Principal	$10,500	$10,500	$10,500	$10,500	0	0
Interest 5%	2,100	1,575	1,050	525	—	—
Parts and Labor	4,000	4,000	3,500	3,000	3,500	3,500
Fuel Cost, 30,000 mi	$11,570	$11,570	$11,570	$11,570	$11,570	$11,570
30,000-mi Cost	0.94	0.92	0.89	0.85	0.50	0.50
Resale	30,000	30,000	30,000	25,000	20,000	20,000
Direct Labor Hours	133	133	117	100	117	117

<u>Average 4 years</u> = Base Year 4 operating cost × 200 vehicles

200 × 0.96¢/mi × 30,000 mi = $5,760,000 is the total cost to own and operate this fleet per year at an average of 4 years

200 × 100 h direct labor = 20,000 h ÷ 1500 h = 13.33 people needed

50% scheduled = 7 bays scheduled work for 7 mechanics

For 6 mechanics doing unscheduled work, 6 × 1.5 bays per mechanic

9 bays for unscheduled work

Total of 16 bays needed 1 shift for 13 mechanics

Resale value of an 8-year mixer @ $30,000 × 25 units per year replacement:

$750,000 received for used vehicles; subtracted from $5,760,000 operating cost = $5,010,000 ÷ 200 vehicles, $25,050 per vehicle per year net cost

$125,000/vehicle new × 25 vehicles = $3,125,000 capital needed per year for a 4-year average age fleet of 200 mixers

Average 6 years = Base Year 6 operating cost × 200 vehicles

200 × 0.94¢/mi × 30,000 mi = $5,640,000 is the total cost to own and operate this fleet per year at an average of 6 years old

200 × 117 h direct labor = 23,400 h ÷ 1500 h = 15.6 people

50% scheduled = 8 bays scheduled work for 8 mechanics

For 8 mechanics doing unscheduled work, 8 × 1.5 bays per mechanic

12 bays unscheduled

Total space = 20 bays.

For the extra 4 bays × 1000 ft^2 each × $125 ft^2 = $500,000 space one-time cost

Resale 12 yr @ $20,000/yr × 16 vehicles = $320,000

$5,640,000 – $320,000 = $5,320,000 net operating cost @ $26,600 per vehicle

$125,000 per vehicle new × 16 vehicles - $2,000,000 capital needed

Average 2 years = Base Year 2 operating cost × 200 vehicles

200 × 0.96¢/mi × 30,000 mi = $5,760,000 is the total cost to own and operate this fleet per year at an average age of 2 years old

200 × 67 h direct labor = 13,400 h ÷ 1500 h = 9 people

50% scheduled = 5 bays scheduled plus 6 bays unscheduled

4 × 1.5 bays per mechanic, 6 bays unscheduled

11 bays

Resale 4 yr @ $37,632 × 50 = $1,881,600

$5,760,000 – $1,881,600 = $3,878,400 = $19,392 per vehicle

$125,000 per vehicle new × 50 vehicles = $6,250,000 capital needed

Summary

Four-year average age:	$5,010,000 @ $25,050 per vehicle Annual replacement $3,125,000	16 bays 13 people
Six-year average age:	$5,320,000 @ $26,600 per vehicle (Additional one-time, four-bay capital cost of $500,000 is needed, or go to two shifts) Annual replacement $1,680,000	20 bays 16 people
Two-year average age:	$3,878,400 @ $19,392 per vehicle Annual replacement $6,250,000	11 bays 9 people

Vehicle Replacement Strategies and Matching Support Systems

The preceding analysis shows us the following:

- 4-year average age cost per vehicle is $25,050

- 6-year average age cost per vehicle is $26,600

- 2-year average age cost per vehicle is $19,392

With this information, we know if we grow older or younger, the difference between staying the same is calculated by subtracting the dollars from the alternatives.

This identifies the capital needs for each of the three alternatives and also the operating dollar needs. Should a decision be made to change direction, we have the capital and operating dollars identified to support that effort.

Proactive efforts are needed for planning to support our mission that addresses our vision.

Chapter 6

Specification Preparation

Fleets and OEMs: Working Together

A fleet manager's primary responsibility is for "safe and economical" vehicle operation. This is accomplished through OEM (original equipment manufacturer) design, plus fleet operation and maintenance. The OEM designs the vehicle for an application-specific environment, with a predictable, cost-effective life cycle based on its correct operation and estimated scheduled and unscheduled maintenance for that application.

Component manufacturers design components to work together to provide reliable service for a predictable time period. An example is an engine that produces 1300 ft-lb at the flywheel/flexplate, which requires a transmission input system to have at least a 1300 lb-ft capacity and a clutch that will handle the transfer of torque between the two components efficiently and reliably.

The OEMs design a vehicle for the real world as defined by the fleet manager. Furthermore, fleet managers come from varied backgrounds, with varied perspectives. Each fleet manager communicates to the OEMs through vehicle specifications, both defined and implied, with language based on their matured perspectives to match a vehicle to its application. The OEM and component suppliers listen to what is said, and they respond with features and benefits keyed to the unique circumstances of each fleet.

Give and take among OEMs, suppliers, and customers helps ensure that customers will reach their expectations, as well as their internal customers' expectations for their equipment utilization and productivity targets. Not only must OEMs meet fleet requirements, they must find ways to increase customers' productivity and safety, and must warrant their products.

A Symbiotic Relationship

OEMs have a tall order. They must meet the challenges set by their customers and remain competitive with their peers. Meanwhile, customers benefit because competition increases available options, extends life cycles, and lengthens and deepens warranties.

OEMs also are called on to provide education. Fleet managers with limited backgrounds such as accounting, shop, customer service, operations, finance, sales, marketing, and executive development bring only part of the total experience package to their positions. They need support from key people within their organizations to carry on the safe and economical mission-vision until they mature in their career positions. OEMs work with these people, supplementing their needs to provide for the needs of their fleets, with the continued vocational needs met by filling in the gaps based on their past performance and experience.

OEMs also extend the versatility and ease of operation of their equipment by sharing other customer developments that result in common standards. Thus, their customer base can take advantage of successful innovations. As the manufacturer gets closer to a standardized vehicle, costs become lower. The more price-competitive a manufacturer becomes, the more units are sold, further reducing costs.

The benefits of improved products also extend to those purchasing used vehicles. Today, the resale market is an active area for fleets, shortening their life-cycle periods without compromising reliability.

Although the resale market is slanted toward the owner operator, it is good for the fleet that has a good maintenance program in place because maintenance cycles for key components have been lengthened. Alternators, starters, clutches, injectors, tires, brakes, and other consumables now require service about 500,000 mi in an over-the-road application; therefore, fleets that dispose of their vehicles prior to 300,000–500,000 mi avoid this maintenance cost. However, the fleet that has a successful maintenance network in place can take advantage of this cost-effective alternative and purchase these used vehicles, take advantage of the lower cost of these depreciated vehicles, and still qualify for the vehicles' warranty program while they assimilate these vehicles into their fleet operation.

Determining Warranties

Fleets purchasing new vehicles must have a good warranty program, which is fleet-designed and spec'd from the menu of items offered by the OEM and its component suppliers for specific applications. For city applications, for example, the warranty time/use periods are shorter than for over-the-road applications because of the different operating characteristics. Stop-go operation, road condition, and environment debris produce different component life cycles.

The OEM provides tested alternatives from which fleet managers may choose, based on their operations. The fleet manager also can ask for customized warranty options, and the OEM will evaluate various test experiences and reply with an offering that often is negotiable between both parties. Again, the cost of innovation is minimized, based on past practices with other fleets and past repair campaign experiences.

Moreover, warranty is an area of competition among OEMs and suppliers, and fleets ultimately benefit—especially larger fleets that receive deeper discounts based on high volume.

Warranty also provides feedback for OEMs. Using the VIN (vehicle identification number) of the chassis, body, and supplier components, the OEM can track the success rate of the performance of each vehicle, based on warranty claims and parts usage. This information is collected and sorted by vehicle and made available to vehicle owners. Fleet managers can obtain the information from their dealers, via the Internet, or by calling their OEMs, and can take advantage of campaigns and warranties before potential problems cause downtime.

Most OEM representatives are proactive with their clients because successful equipment operation, as judged by operator and mechanic acceptance, means return business. A proactive approach benefits all parties—manufacturers, fleet managers, drivers, mechanics, and internal and external customers.

Training/Information

OEMs provide high-quality training for fleet staffs, including drivers and technicians. As new components become available, the knowledge of how

133

those components work, what they do, and what to measure makes for extended life and reliable service. Electronic control units, antilock brakes, and tractor-trailer communication systems are among the newer components that require upgraded training. Each year, OEMs offer the information we need for intelligent product selection and operation.

Consumables such as tires, alternators, brakes, and starters are developed, based on customer needs and the past experience of OEMs. This offers fleets longer life cycles of these parts in their applications—increasing vehicle usage and lowering overall costs. Training is available for these consumables by OEMs, so fleets can better understand design objectives and perform proper maintenance to achieve longer component life cycles, reduce costs, and improve uptime.

Figure 6-1 shows the costs of running 1,000,000 mi for a 1975 tractor versus a 1995 tractor. The cost decrease for newer equipment is attributed to technological innovation.

	Old vs. New Tractor Maintenance Cost Differences over 1,000,000 mi	
1975		**1995**
3–4	Engine Rebuilds	0
2	Transmission Rebuilds	0
2	Rear Axle Rebuilds	0
4–5	Clutch Replacements	1
8	Brake Jobs	3
83	Oil Changes	41
10	Sets of Tires	4
4	Miles per Gallon	8
$150,000/$165,000		$65,000

Figure 6-1
Costs of Running 1,000,000 mi for a
1975 Tractor Versus a 1995 Tractor
(Courtesy of Ryder Transportation Services, Miami, FL)

Reduced Maintenance Costs

We all should have cost information systems. If we do not know cost, we cannot manage efficiently. Management information systems provide component, total vehicle, and average class-of-vehicle costs. This enables us to inspect each group of vehicles for information so we can ask intelligent questions to improve performance and reduce cost without sacrificing reliability and life-cycle periods.

Based on OEMs' global competitive efforts, we have seen costs change in favor of fleets (see Figure 6-1). No more are there gradual, increasing, annual maintenance expenses. Due to increased component life cycles and reliability, and extended consumable life cycles, equipment maintenance should show lower costs per mile for longer periods.

When a vehicle enters the fleet and we weed out warrantable issues, our costs level out rather than gradually rising as in the past (Figure 6-14). This affords us longer life at less cost per mile, per hour, and per product delivered.

Each fleet manager in the late 1980s to early 1990s was charged internally to reduce cost and increase reliability, life cycles, and productivity. The OEMs and fleets accepted the challenge together, defined specific expectations, modeled costs through historical information, and set out to achieve their goals. Most of us listed and prioritized our total annual maintenance costs by component, then by vehicle-class cost, and then reviewed our highest cost areas. Fleets then challenged OEMs and suppliers to design components that would reduce costs in our applications.

What evolved was a list of cost areas in specific applications—preventive maintenance, brakes, lights, suspensions, engines, tires, steering—which focused on the appropriate components that needed attention by total dollars spent.

In other words, we are trying to receive the best bang for our bucks. When the highest-cost areas are handled, we can move to the next-highest components—air conditioning, cooling systems, instrument switches, and so forth.

The Big Picture

With environmental cleansing today, alternative fuels and synthetic replacements for petroleum products have a higher profile, based on government support and funding for OEMs to use in research and development. Likewise, it is common to see fleets volunteer as beta sites for real-world testing to validate findings.

The OEM will share with us repair standards and offer training to support productivity efforts internally and help estimate repair costs.

Estimating vehicle damage and troubleshooting technical and unusual problems through OEM dealers is an invaluable resource and network for problem solving because the OEMs have the latest test equipment and appropriate staffs to support this effort.

The manufacturers can simulate vehicle design changes through CAD (computer-assisted design) systems to eliminate trial-and-error design problems. This gives fleets opportunities to explore the benefits of alternate specifications such as wheelbase and weight-distribution changes, speed and shift point changes, and cab alternatives for productivity upgrades.

Overall, the resources offered by OEMs benefit our applications and the economical performance of our companies. We all play on the same team, and cooperation benefits all of us, including our internal and external customers.

Specification Differences

In the preparation of a vehicle specification, you are concerned with two approaches. These are a "technical specification" and a "functional specification."

Technical Specification

A technical specification quantifies an item to the most parturient detail and impacts vehicle design. It also puts the responsibility for liability directly on the vehicle specification writers and their company.

For example,

> "...The hydraulic line will be 3/8 in. inside diameter, made of stainless steel and carry a 1000 psi pressure rating. The input line will be 4 ft long, painted green with no more than 2 90° bends..."

Here, one is accepting the liability by specifying pressure, size, routing, etc.

Functional Specification

A functional specification would be a definition of a task, such as, "...We need the boom to rotate 360° and pick up a maximum of 1000 lb at 45° and 20 ft from the chassis in any position...."

Here the task is defined, and the manufacturer determines the components needed to meet the customer's need and thus assumes the liability.

Technical Specification Benefit

The benefit of the technical specification process is that you can go out for bids with the goal of accepting the lowest-priced bid, as long as the manufacturer meets your technical specified components.

Functional Specification Benefit

With a functional specification, other items must be taken into consideration in the evaluation of the bid:

- Technical requirements met. Over- and underspecified components measured to verify compliance. Dollar credit given to overspecified components and deductions for underspecified items.

- Quality and reliability of each bidder's response to ensure that its designs are made to last with a minimum of support maintenance. Should one unit cost more to maintain over its lifetime than another? This information should be added into the evaluation of the bid using historical maintenance data projected into the future cost of the maintenance of that model.

Vendor A requires an inspection of Item 1 three times per year because of its design and assembly costing $500 per year more to maintain than Vendor B, C, D, & E's units, which require only an annual inspection of Item 1.

The functional specification preparation process will be pursued because that is the process most widely used by the majority of transportation organizations.

Sample Functional Specification

A functional specification should note components by brand name and add in the general introduction the phrase "or equivalent."

Sample Tractor Specifications

Freightliner Condor COE 2-axle non-sleeper 114-in. wheelbase truck, or equivalent

Engine–Cummins ISM–Caterpillar CFE–290 hp, or equivalent

Front Axle–Meritor non-driving single-axle, 12,000 lb, or equivalent

Front Suspension–Single taperleaf, 12,000 lb, or equivalent

Allison MD series automatic transmission, or equivalent

Meritor single, rear axle with outboard mounted hub and drum assembly– 4.11 ratio, or equivalent

Front engine transmission PTO, or equivalent

Chassis Specifications
Conventional Chassis 15,000-lb Payload for a 20-yd Packer
Rear Loader 750 lb/cubic yd

	Comply	
	Yes	No
Conventional Chassis, new unit, never registered	____	____
Diesel engine—minimum 250 hp, electronic, Interstate 55 mph loaded	____	____

	Comply	
	Yes	No
Automatic transmission, five-speed, deep reduction first gear, refuse, plow options, cooler, backup alarm, PTO pump	____	____
Single-axle rear, maximum legal weight 21,000 lb, 23,000 susp.	____	____
Single-axle front, workable weight distribution 10,000 lb, 10,000 susp.	____	____
Maximum front and rear braking system, air, antilock, maximum air compressor, auto slack adjusting	____	____
Shortest workable wheelbase to distribute 31,000 GCVW	____	____
Front and rear axle bearings liquid lubrication	____	____
Standard interior, tan/gray	____	____
Basic exterior standard cab, white, air horn	____	____
Maximum cold-cranking 12-V batteries, maintenance free	____	____
100-gal fuel tank, diesel	____	____
Seats, driver air suspension, passenger bench	____	____
Mirrors, standard West Coast, stainless, retractable	____	____
Tilt hood with integral fenders and liners	____	____
Standard cooling system, silicon hoses, viscous fan clutch	____	____
Vertical exhaust system with shields	____	____
Exhaust engine brake	____	____
110-V plug-in electric cooling system heater, engine-mounted plow attachments, tow hooks	____	____
Safety equipment, triangles, 5-lb fire extinguisher	____	____
Standard tire12Rx22.5, disc wheels, pilot hub mounted steer front, lug rear thread tubeless	____	____
125-amp alternator	____	____
All service and parts manuals will be supplied on delivery of the vehicle chassis in hard copy and CD-ROM.	____	____
A full one-year warranty, full parts and labor, bumper to bumper is required.	____	____
A full height distribution of chassis and mounted	____	____

Chassis Specifications
Low Entry Cab Over Chassis 15,000-lb Payload
for a 20-yd Packer
Rear Loader 750 lb/cubic yd

	Comply	
	Yes	No
<u>Low Entry Cab Over Chassis</u>, new unit, never registered	____	____
Diesel engine, minimum 250 hp, electronic, Interstate 55 mph loaded	____	____
Automatic transmission, five-speed, deep reduction first gear, refuse, plow options, cooler, backup alarm, PTO pump	____	____
Single-axle rear, maximum legal weight 21,000 lb, 23,000 axle, 23,000 susp.	____	____
Single-axle front, workable weight distribution 12,000 lb, 12,000 susp.	____	____
Maximum front and rear braking system, air, antilock, maximum air compressor, auto slack adjusters	____	____
Shortest workable wheelbase to distribute 33,000 GCVW front and rear axle bearings liquid lubrication	____	____
Standard interior, tan/gray	____	____
Basic exterior standard cab, white, air horn	____	____
Maximum cold-cranking 12-V batteries, maintenance free	____	____
100-gal fuel tank, diesel	____	____
Seats, driver air suspension, passenger bench	____	____
Mirrors, standard West Coast, stainless, retractable	____	____
Tilt cab, hydraulic close-open	____	____
Standard cooling system, silicon hoses, viscous fan clutch	____	____
Vertical exhaust system with shields	____	____
Exhaust engine brake	____	____
110-V plug-in electric cooling system heater, engine-mounted plow attachments, tow hooks	____	____
Safety equipment, triangles, 5-lb fire extinguisher	____	____
Standard tire 12Rx22.5, disc wheels, pilot hub mounted	____	____
125-amp alternator	____	____
All service and parts manuals will be supplied on delivery of the vehicle chassis in hard copy and CD-ROM.	____	____
A full one-year warranty, full parts and labor, bumper to bumper is required.	____	____
A full height distribution of chassis and mounted	____	____

20-Cubic-Yard Rear Loader Packer Unit Specifications

	Comply	
	Yes	No
Maximum 20 cubic yd capacity exclusive of the 2–3-yd hopper	___	___
Compaction: 750 lb/cubic yd	___	___
Must meet all ANSI safety standards	___	___
Body weight estimated on or about 10,000 lb empty	___	___
Must be close to rectangular in shape	___	___
Must specify maximum width, height from ground, length front to back	___	___
Must provide certified weight distribution, empty and loaded	___	___
Skid-resistant steps and handrails	___	___
Amber strobe lights mounted on rear in addition to rear tail marker, stop, directional, spotlights; FMVSS#108 lights and reflective midbody turn signals	___	___
Must have conspicuity tape on sides and rear	___	___
Must have buzzer system to cab from tailgate both sides	___	___
Packer controls positive lever type with manual throttle both sides	___	___
Ejection and tailgate lift controls positive lever type with manually actuated throttle both sides	___	___
Detail roof, side construction, features and benefits	___	___
Detail body reinforcement construction, features and benefits	___	___
Detail trough, floor sheets, braces, supports, access doors/panels, features and benefits	___	___
Detail body frame features and benefits	___	___
Detail inside and exterior body dimensions, features and benefits	___	___
Detail hopper construction, operation features and benefits	___	___
Detail packing mechanism features and benefits, plates, cylinders, operation	___	___
Detail tailgate construction, operation features and benefits	___	___
Detail ejection systems, operation features and benefits	___	___
Detail hydraulic system operations, maintenance features and benefits	___	___
Detail operating controls, features and operations	___	___
Lighting and electrical system must meet FMVSS#108 regulations, sealed harnesses, hopper light, backup alarm wired if tailgate is not closed and vehicle is in reverse, sealed junction boxes	___	___

20-Cubic-Yard Rear Loader Packer Unit Specifications (cont.)

	Comply	
	Yes	No
Paint, shot-blasted, clean surface, steam-cleaned, soaped, two-part epoxy primer, two-part polyurethane top coat, forest green baked at 200°F	____	____
American Goods and Form Products NJSA40: 11-1 et seq.	____	____
12,000-lb overhead winch mounted top of tailgate with lock-type container attachment	____	____
Provide leaf chute 1/4 in. × 1-1/2 in. × 1-1/4 in. angle iron 96-in. opening plus lifter with conventional chassis only	____	____
Provide a list of parts and supplies needed for one year of maintenance.	____	____
Provide a detailed sequence PM (Preventive Maintenance) program for chassis and body.	____	____
All parts and warranty service must be within 24 hours of notification.	____	____

Conventional Chassis Weight Estimation
FA - 10,000 lb capacity   6,445 lb
RA - 23,000 lb capacity   4,141 lb
                                                    10,586 lb total

Cab Over Low Entry
FA - 12,000 lb capacity   6,945 lb
RA - 23,000 lb capacity   3,641 lb
                                                    10,586 lb total

Body, 20-yd packer, 2–3-yd hopper   9,900 lb

Payload, 750 lb/yd × 20 yd   15,000 lb

Chassis   10,600 lb
Body   10,000 lb
Payload   15,000 lb
                                                    35,600 lb

Conventional

10,000 lb front axle
21,000 lb rear axle
31,000 lb toal GCVW

Cab Over, Low Entry

12,000 lb front axle
20,000 lb rear axle
33,000 lb total GCVW

Upon issuance of a purchase order number, the successful bidder will make available two rear loader packer units and chassis for use of the company until the units requested are assembled, delivered, and accepted by the company for payment.

At the successful bidder's cost, three people from the company will be transported to the assembly site of the finished unit to inspect, operate, and review issues at the prepaint stage of the packer unit mounted on the chassis.

At this point, all criticisms, observations, and comments of the company personnel will be discussed from a handwritten document and corrective action items recorded, typed by the vendor, and distributed to all present, so that upon delivery, all corrections will be made so the company fleet services unit will inspect and in-service unit to validate turnkey condition of the vehicle, or invoice the liquidated damage terms and conditions.

The objective is to create an environment of competition without "wiring" a specification limited for one make model. If a specific brand is what you want, you should stimulate all vendors of that brand to bid or propose an equal or equivalent item.

Most heavy vehicles produced are a mixture of many sub-components supplied and assembled by a manufacturer. Light and medium vehicle components are produced in general by a manufacturer coordinating subassemblies under its name. You may be comfortable with the assembly process, but you'd be more comfortable with a heavy manufacturer that would make available more components than one specific brand. Competition provides your best value.

If a dealer is local and gives convenient, reliable, and cost-effective service, you should write other area dealers of the same brand to participate in the bidding process and measure their potential service efficiency. You can compare each vendor line item by line item to resolve unit cost variances, not only in unit price but in component costs per line item.

Vehicles of different manufacturers have similar components, such as, engines, transmission, brakes, and clutches. This makes multi-brand component vehicle manufacturers a good common source because they would have service available for the same brand engines, transmission, brakes, clutches and so forth to service.

Another consideration to include in your specifications is the need for each order of vehicles to include spare parts for your shop inventory to be included with the vehicle delivery.

This prepares your support organizations to have available fast-moving items to service your vehicles in a timely manner.

Many times when transportation orders fast-moving items for new vehicle service, (e.g., brakes, air filters, oil filters, fuel filters, belts, and fittings), they are being prioritized for the manufacturing process and are not available from suppliers until fleets start asking for them.

Another item to include in the specifications is "birth certificate" information for your record keeping. A parts line sheet (Figures 6-2 and 6-3) is usually included with a chassis delivery. It lists the OEM parts numbers for your use in parts ordering. These line sheets should be made available for vehicle bodies and mounted equipment such as dump bodies, cement mixer barrels, aerial devices, trailers, lift gate and platforms, and so forth.

In the day-to-day maintenance of your vehicles, most of your vehicle maintenance records are computerized. This allows for more accurate cost systems and inventory systems, but it requires accurate and timely input. Figures 6-4 and 6-5 are examples of information needed to be input upon receipt of your vehicles, and it is desirable on your part to require the manufacturers to fill out these forms completely for your use immediately prior to the vehicle being delivered.

When the vehicle is delivered, you can verify the information on Figures 6-4 and 6-5 and input the data for vehicle service support information.

Prior to assembling the vehicle specification, the cost for your present vehicles should be reviewed to identify high-frequency and high-dollar areas of cost.

F8000	1FDYK87U3HV A63706		9754A	13154

40 7125	F O R D M O T O R C O M P A N Y	TRUCK SPECIFICATIONS LIST

Dealer Order # 87 T00705   Series F8000   Wheelbase 178   Serial # 1FDYK87U3HV A63706
Page

0S0 #13-4915	Type 13154	1 9754A

Code Qty Part Number	TSL Description	Part Discription	Part 0S0 Option

Wheels, Hubs and Drums

0106	EGHT	1102 AA	Hub & Drum (Frt)	Alt W/E 4HTGA
0106	EGH	1103 AA	Hub & Drum (Frt)	Alt W/E4HTGA
0112	EGHT	1113 MA	Hub & Drum (RR) RH	8819A Repl E4HT-ACE
0115	EGHT	1114 MA	Hub & Drum (RR) LH	5819A Repl E4HT-ACB
0118	2.E7HT	1190 AA	Seal (Frt Wheel)	C421 Ret. Asy. Ft. Whl
0121	2.E5HT	1175 GA	Seal (Rear Wheel)	C421 Seal Asy.
0124	7.E7HT	1015 DA	Wheels	F5344 Wheel 27404N

Figure 6-2
Chassis Line Sheet

This will put the focus on components that need reviewing to bring to your attention new components that will be, although higher in initial cost, lower in overall cost.

A simplified example: If repairs are made to 400-amp cold-cranking batteries and 70-amp alternators on a two-year basis, maybe lighting and accessory need is 75 amp peak, 65 amp average, causing the system to overwork and fail prematurely.

It would be in the best interest of the company to consider specifying and ordering a higher-capacity battery and higher-output alternator. The manufacturer will work with you in reviewing the needs of your company and the capacity of the components available. The amperage draw can be totaled, and a decision can be made on the upgraded component to use.

Component	Mfr.	Part/Model NC.	Ser. No.
PYLON MAST ASSEMBLY			
Lower Control Valve	Dukes	818-0101-75	2727DF
Interlock Valve	N/A	N/A	N/A
Lower Room Cylinder	BREW	172-003-44	2304-11-07
Holding Valves	SUN	818-0600-07	F706-15W-40
Return Filter	Lenz	802-0027-00	N/A
Hydraulic Tank	Leadom's	178-0296-44	N/A
Shutoff Valve	Sterling	818-1000-02	N/A
Rotation Speed Reducer	Gior Prd.	172-0002-33	FP-00009
Flow Regulator	N/A	N/A	N/A
Rotation Hyd. Motor	Charlynn	817-0200-14	N/A
Rotation Coupling	Hydromotn.	028-1015-43	N/A
Rot. Relief & Holding Valve	Racine	818-0300-07	N/A
Electrical Collector Assy.	Hydromotn.	002-0206-33	1515
Decals	Beware	006-0085-23	N/A
Internal Tank Strainer	Oil Filtr.	802-0009-00	N/A

Figure 6-3
Mounted Equipment Line Sheet

Overspecification is not desirable because of the increased expense of the components and also the horsepower consumed from the engine reducing efficiency and miles per gallon.

A manufacturing projected fuel use estimates for each 20 hp needed to power accessories, an additional gallon per hour would be used, which is an estimated 15% reduction in fuel economy. It is desirable to minimize the use of components not needed.

VEHICLE BODY INVENTORY INFORMATION   BI

Vehicle Number _____   Body Number _____

Body Year _____   Body Manufacturer _____

Body Model No. _____   Body Model Des. _____

Body Weight _____

Body Own/Lease _____   Body Lessor P.O. No. _____

Body Serial No. _____

Body Rebuilt (Y/N) _____

Body Winch Model No. _____   Body Winch Model Desc. _____

Body Winch Serial No. _____

VEHICLE MTD EQUIPMENT INVENTORY INFORMATION   MI

Vehicle Number _____   Mtd Equip Number _____

Year _____   Mfg. _____

Model No. _____   Model Desc. _____

Weight _____   Lessor P.O. No. _____

Own/Lease _____   Serial No. _____

Engine Mfg. _____   Engine Model _____

Engine Ser. No. _____

Prim. Fuel Type _____   Prim. Fuel Cap. _____

Radiator Cap _____

Transmission MDL _____   Trans. Mfg. _____

Trans. Serial No. _____

Winch Model No. _____   Winch Model Descrip. _____

Winch Serial No. _____

Mainframe Ser. No. _____   Rebuilt (Y/N) _____

VEHICLE SNOWPLOW INVENTORY INFORMATION   SI

Vehicle Number _____

Snowplow Number _____

Snowplow Year _____   Snowplow Mfg. _____

Snowplow Model No. _____   Snowplow Model Desc. _____

Snowplow Serial No. _____

Snowplow Own/Lease _____   Snowplow Lessor P.O. No. _____

Figure 6-4
New Vehicle Body Inventory Birth Certificate Information

VEHICLE WEIGHT INFORMATION   WI

Vehicle Number _____

Wheel Base _____   Number of Axles _____

Drive Axle _____   Front Axle Weight _____

Rear Axle Weight _____   Taxable Weight _____

Empty Weight _____   Chassis Weight _____

GVWR _____   GCVWR _____

VEHICLE DRIVE INFORMATION   DR

Vehicle Number _____

Primary Fuel Type _____   Primary Fuel Capacity _____

Secondary Fuel Type _____ Secondary Fuel Capacity _____

Engine Serial Number _____

Engine Model Number _____

Transmission Mfg. _____   Transmission Model _____

Transmission Ser. No. _____ Brakes _____

Front Tire _____   Rear Tire _____

Clutch Size _____   Clutch Type _____

VEHICLE MISCELLANEOUS INFORMATION MS

Vehicle Number _____

Ignition Key Number _____

Trunk Key Number _____

Enter a Y or N for the Following Items

Backup Warning Signal _____ Air Conditioner _____

Jib Boom _____   Outrigger _____

Tool Outlet _____   Generator 120V _____

Snow Plow _____   Power tailgate _____

Salt Spreader _____   Rustproofed _____

PM SCHEDULE INFORMATION PM

Vehicle Yearly Insp. _____   Vehicle 3 Year Insp. _____

Vehicle PM Mo 1 _____   Vehicle PM Mo 2 _____

Vehicle PM Mo 3 _____   Vehicle PM Mo 4 _____

Vehicle PMC 1 _____   Vehicle PM Mo 6 _____

Figure 6-5
New Vehicle Chassis Inventory Birth Certificate Information

New electrical controls for gas and diesel efficiencies cause parasitic drag when the unit is shut down. This amperage is drawn from the battery, reducing battery storage capacity, and should be considered in the battery component specifications.

Tractor-Trailer Power Usage*2

Device	Amps	Hrs.	Amp Hrs.
Key on	2.0	10.0	20.0
Starter	150.0	0.1	15.0
Wipers	7.5	10.0	75.0
Headlights	8.0	10.0	80.0
Panel lights	3.0	10.0	30.0
Tail lights	1.5	10.0	15.0
Marker lights	12.0	10.0	120.0
Brake lights	8.0	0.25	2.0
Turn signals	5.0	0.15	0.75
Engine electronics	6.25	10.0	62.5
Heater fan	15.0	10.0	150.0
Heated mirrors	5.0	10.0	50.0
CB radio	1.5	10.0	15.0
AM/FM radio	1.0	10.0	10.0
Radar detector	0.5	10.0	5.0
Subtotal			650.25 amp hrs
+25% safety factor	91.25 amps		162.56 amp hrs
Total			812.81 amp hrs

* Calculated for 10-hour shift on a winter night.

---

2. *Heavy Duty Trucking Magazine*, October 1989.

Frequency of Repairs

Further analysis of Figure 6-6 reviews frequencies of 50 or more occurrences per month in the day-to-day operation.

The question of excesses here is gray. The category of PMA (preventive maintenance inspection) at 126 incidents is good, as long as you are on schedule. Road call at 143 incidents is perceived as not so good.

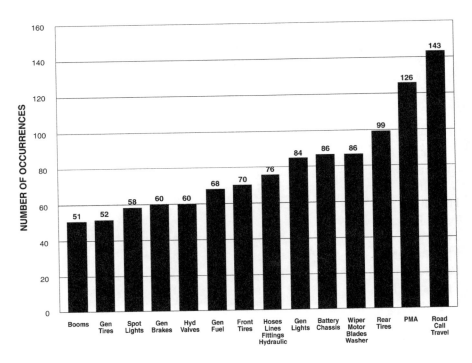

Figure 6-6
Road Call Frequencies

Front, rear, and general tires at 221 incidents is perceived as not good, but it depends on your operating on or off the road and how many vehicles you have. A fleet of 6,000 heavy vehicles would be better off than 1,000 vehicles in this area.

Battery, lights, and spotlights at 228 incidents is significant overage.

Windshield wipers, motors, and blades at 86 is questionable.

Hydraulics, valves, lines, and fittings at 136 incidents is questionable also.

This is not enough information with which to make a decision. Rather, it is intended only to initiate a question and a fact-finding process to analyze the situation for better grade of components or maintenance to reduce the frequency of repairs.

As far as an alternator need here, with 91.25 amps generally needed, a 125-amp alternator would be a better choice than a 75-amp or a 90-amp unit. The 75- or 90-amp unit would be underpowered.

Battery cold-cranking capacity should exceed 800 cold-cranking amp hours to provide the reserve necessary in this application. Anything less is engineered to fail.

Repair Code Costs

The next area to review is repair costs, the exception of $10,000 per month. Similar frequency, it skims the top few costs to review and compare, so one can ask a more productive question (Figure 6-7) and look for similarities in frequency (Figure 6-6) usages.

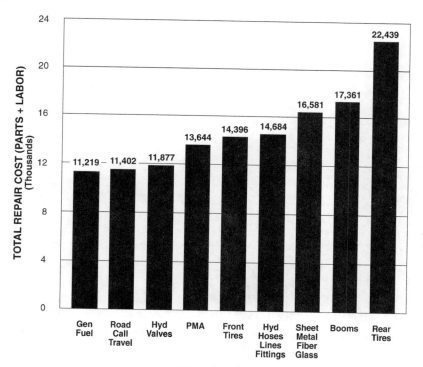

Figure 6-7
Repair Code Costs

Here, tires, road calls, hydraulic hoses, lines, fittings, valves, PMA, and fuel systems are highlighted in both cost and frequency. This initiates an inquiry into the circumstances and through fact-finding and analysis will provide a basis for the possible need for a better grade or capacity of components to reduce repairs in these areas also.

Then the increased cost of components can be taken and the decrease of the repair costs estimated to see if it is cost-effective to install the newer, more costly component.

Road Calls

Road calls are a red flag. Not only is it more expensive to fix a vehicle on the road, but cargo, crew, and passenger delays exacerbate unscheduled expenses.

On Figure 6-8, hydraulic, electrical, and chassis road call frequencies complement information from Figures 6-6 and 6-7 in a quest to reduce costs.

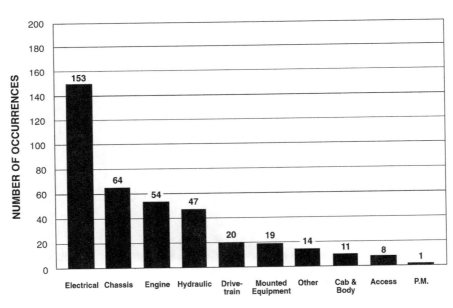

Figure 6-8
Road Call Repair Code Frequency

The one PM road call is probably bad data that should not have been reported. You should be careful in the interpretation of your information systems to reduce false starts. This is an area of concern in fact-finding and analysis. Comparison of data demands an experienced person to verify the message.

Vehicle Cost Due to Age

Figure 6-9 shows maintenance trends. These tend to increase as a vehicle ages.

Figure 6-10 is a graph of a straight truck maintenance cost per 10,000-mi increments.

Figure 6-11 shows the top 8 to 10 component costs during each 10,000-mi maintenance interval. It shows the components on which to focus to see if, by

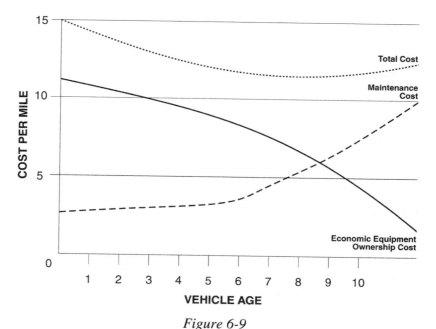

Figure 6-9
Estimated Vehicle Cost Per Year
(Compliments Fleet Equipment; John Larkin; Alex Brown & Sons, Inc.)

153

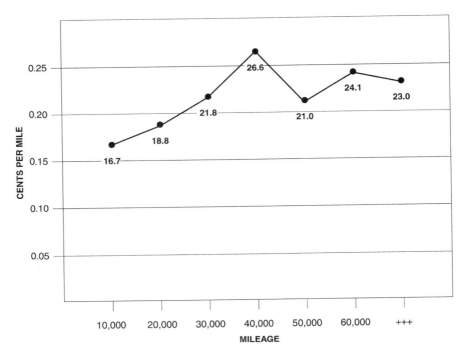

Figure 6-10
Heavy Straight Truck 40,000 lb GVW–Maintenance History

increasing their capacity, resultant life can be extended, reducing maintenance costs overall.

How to Properly Spec Your Vehicles

A written vehicle specification details the service in which you expect your vehicle to function. This functional specification would say, "I want a 12-yd dump truck to travel 55 mph loaded up a 6° incline while towing a 10,000-lb trailer with a backhoe on it."

With a functional specification, the manufacturer assumes the liability and accountability to produce a safe and functional vehicle.

0-10,000	10-20,000	20-30,000	30-40,000	40-50,000	50-60,000	60,000+++
PM..........17,000	Tires 13,200	Tires 19,000	Tires 20,000	Tires 17,000	Tires 12,100	Power ...35,000
Frame 46,000	Cool 3,800	Hyd.Cyl 4,400	Light 3,900	Light 3,100	Clutch 4,300	Cool ...11,000
Tires 8,500	Brakes 10,000	Brakes 10,000	Brakes 10,000	Brakes 7,500	Brakes 6,800	Tires 34,000
Brakes.......8,500	PM...........8,700	PM............6,000	PM..........6,000	Power 6,400	PM...........5,200	Brakes ..27,000
Light 8,348	Light 7,340	Light 4,400	Fuel 4,200	PM............6,000	Light 4,500	PM.........1,200
Cool 4,600	Hyd.Cyl 3,500	Clutch 3,700	Susp 3,200	Fuel 2,800	Cool 3,600	Light 10,000
Hyd.Cyl 4,500	Steer 3,349	Cool 3,600	Clutch 2,800	Cool 2,000	Trans 3,300	Exhaust ..9,000
...16.7¢/m...	...18.8¢/m...	...21.8¢/m...	...26.6¢/m...	...21.0¢/m...	...24.1¢/m...	...23.0¢/m...

Figure 6-11
Heavy Straight Truck 40,000 lb GVW

We would specify the type of engine, driveline cab, and interior based on the selection available from the manufacturer. After each line item, we would say "or equal or equivalent" to invite multiple bidders. The more bids on one solicitation, the more competition and the better the price.

We must look at the vehicle being replaced to see what maintenance and operating issues it has. Is the cab the correct configuration? Does the body hold the cargo for which it is designed? We should weigh the present vehicle loaded to see if the weight distribution empty and loaded is safe and proper on each axle and compare the old axle specifications to the new specifications to determine if any changes are needed.

A single axle is rated up to 21,000 lb, and a tandem axle up to 34,000 lb. Axles and suspensions are manufactured for lower and higher weights, but weight laws require that your vehicle must be in minimum and maximum compliance. As you request and specify a vehicle, the weight distribution should meet federal and local gross vehicle weights, gross vehicle combination weights, and axle weight requirements.

When a vehicle is designed, the manufacturer will weight the vehicle in its design stage "CAD" (computer-assisted design) and when it is built "CAM" (computer-assisted manufacturing). The manufacturer also will conform to the design that will be safe and meet regulatory laws. It is the manufacturer's accountability and responsibility to manufacture a safe and efficient vehicle.

Thus, when we sign an agreement to purchase a vehicle, we should note the weight distribution and require a certified scale weight slip upon delivery to enable us to validate confirmation of the "CAD-CAM" design specification.

In addition, we must calculate the speedability and gradeability of the vehicle. Speedability is the shift points of the vehicle to be between the maximum engine torque and maximum horsepower rpm of the engine. If the shift points fall below the maximum torque when shifted, then we must increase the engine horsepower/torque capability or add more shift points. The manufacturer is responsible on a functional specification to assure we have efficient shift points. Any marginal points should be identified, discussed, and resolved prior to agreement on the vehicle for manufacture and delivery, especially if there is mounted equipment on the chassis such as a dump body, aerial device, refuse packer unit, cement mixer, and so forth.

Gradeability is the ability of a vehicle to ascend a specific degree grade fully loaded in the lowest forward transmission and rear axle gear. This calculation is important to ensure that the vehicle is able to work at its maximum load safely and efficiently.

When the chassis, body, and mounted equipment specification is drafted, each vendor that is supplying its section should attend a pre-bid meeting to review our specifications. The vendors should make recommendations based on their observations, with all concerned parties together to hear the same words at the same time, our end user and ourselves. This ensures that all questions are answered and any problems we have experienced with our old trucks, the one that is being replaced, are corrected to everyone's satisfaction.

With the vehicle specification written, the next step is to solicit price bids on the vehicle. We want to create a competitive environment to stimulate the lowest price for our functional specification.

Whether we are committed to one make of vehicle or we are open to various makes, we should communicate and solicit our specification to as many bidders as possible within our service territory.

To pursue this end, we would ask all vendors to attend a pre-bid conference. This allows us to address any questions in the full presence of all bidders. We would circulate the functional specification prior to the meeting and review, line item by line item, the technical items required. The more complex the vehicle, the more time is needed to clarify items and coordinate the chassis, body, and mounted equipment installation process.

With the vehicle chassis, body, and mounted equipment specification presented, all manufacturers and/or suppliers must be present. A matter of concern is the chassis construction time because the body manufacturer must order, assemble, and mount the body. The final installation of the mounted equipment should be scheduled after or during the body specification review. Body and chassis subframes and the assembly process must be reviewed and coordinated to ensure the proper fit and functionality.

If the mounted equipment is to lift 900 lb at a horizontal 180° angle, the chassis and body frame, mounted equipment subframe, and the mounted equipment mounting process must be reviewed technically. Will the stresses be acceptable by the chassis frame, body frame, and mounted equipment subframe to support the weight lifted?

With dump bodies, cement mixers, and plows, the proper chassis strength and construction are required for long life. With a group pre-bid meeting, the customer will receive input from all present at that time, including chassis and body manufacturers, to ensure that the functional specifications are valid.

Timetable of Deliveries

We must define the format of the process and the timetable of the process. For example,

"We request that you send the quote to us in one envelope, addressing line item by line item each component supplied by brand name or equivalent; and a second envelope with the prices defined. Our process will be to evaluate the bid technically for qualification separately from the price to fairly evaluate your offer. If the bid meets the requirements, we will combine that evaluation with the quality and business evaluation of your company and the price you quote, awarding the order to the best-qualified complete package submitted by your organization."

We want to further define our expectations of delivery of the order and our inspection of the work as it progresses.

For example,

"We will inspect the first chassis when it comes off the assembly line for its conformance to specification prior to being delivered to the body and/or mounted equipment manufacturer no later than 90 days from the award of the order. We will pay 70% of the chassis manufacture cost 30 days prior to completion of the entire assembled unit or 60 days after we approve the chassis. At that time, we expect the manufacturer's statement of origin, so we can register the entire unit to be put on the road when delivered to our property.

"We expect to inspect the body and/or mounted equipment assembled and operating but not painted. Upon approval of the prepaint inspections, the first unit will be corrected and painted for our final inspection. The vendor will pay for our out-of-pocket expenses at that time and for future visits. Our time is to be priced at $50 per hour, or whatever is appropriate, from departure of the home office until returning, plus out-of-pocket and/or agreed upon payment."

This point must be clarified to ensure that the customer is not used as an inspector. We intend and want the manufacturer to use our specification as a guide to inspect its work to our standards.

Notification of Inspection Dates

We must define the notification process of the prepaint inspection date three weeks prior, and the delivery date must be defined so that the customer can prepare for the vehicle acceptance process. When the vehicle is delivered, we want to inspect it, perform an acceptance preventive maintenance procedure, and decal and install items for service such as fire extinguishers, transfer of the stock, and so forth. After the vehicle is prepared for in-service, we should quantify an acceptance time for an in-service check to ensure that all accessories work properly.

For example,

"The unit will be put into service two weeks after delivered on-site to the customer, and the shakedown period will be 30 working days after in-service. After the 10-day in-service and 30-day shakedown period, the unit, if satisfactory, will be approved for payment. A check will be generated 20 working days after acceptance, which is 60 days after delivery to the customer's property."

Should the unit be troublesome and the problems not corrected during the 30 working-day shakedown period, we must quantify the action taken.

For example,

"Should the vehicle not be approved for service 40 days after delivery, payment will not be authorized. If the unit is not corrected 30 days after the 30-day shakedown period, which is 70 days after delivery, a $3,000 per month penalty (or whatever penalty dollar amount is appropriate to offset the increased expense the old unit is costing the company to keep it in service) will be subtracted from the involved amount (liquidated damages)."

Warranties

Warranty expectations should be defined. Standard warranties are supplied by the manufacturer, or you can quote your own to support your own needs (e.g., on-line communication versus paper transactions).

For example,

"All work will be warranted for two full years from date of completion and acceptance by the customer."

Another area of concern is vehicle latent defects. These are design failures due to poor workmanship or poor materials that could not ordinarily be found during a normal inspection process by the manufacturer or the subcontractor supplying the body, mounted equipment, or components that compose them. Statutory law notes any latent defects are payable for 48 months from the date of acceptance for payment. You might want to quote 60 months to exceed the state requirements to cover failures, as a means of keeping the door open for our valid claims process should it be needed because our expectations exceed the state's and the vendor agrees to this item. An example is chassis frames; they last more than 20 years. It is implied that if a frame fails before 20 years or after 48 months, the 60-month latent defect clause opens the door to negotiate an implied warranty settlement.

Costs to Vehicle Bids

These previous items could cost us extra in the price of the vehicle. The vendors must understand the terms and conditions of our bid requirements to enable them to bid accordingly. With the competitors established, each vendor will carefully tender its proposal and moderate extra cost inclusions.

If we have had success with vendor delivery and small, medium, and/or heavy-duty vehicles with no in-service problems, we should be cautious about additional requirements. In most cases, due to people processes and material problems by the manufacturers (similar to our people process and material problems), we should quote and define our expectations to ensure a timely and cost-effective in-service process.

The more detail we include in our specifications, the more room for a misunderstanding to take place. Thus, a pre-bid request for a proposal meeting should take place to ensure full understanding of customer requirements. It is fair to both the manufacturer and the customer.

Automated Fleet Management Information Systems

Fleet management information can be gathered many ways. It can be acquired manually—through work orders, preventive maintenance record sheets, fuel records, parts records, electronic controls on vehicles, onboards, and pre- and post-trip inspections of equipment. It can be gathered orally, when fleet people talk to each other—drivers, equipment operators, mechanics, technicians, semi-skilled laborers, parts people, clerical personnel, data entry clerks, head mechanics, and supervisors and managers. Increasingly, information can be gathered electronically.

When it comes to information gathering, language can be arbitrary and deceptive, particularly when it is "technobabble," jargon, or slang. That contributes to errors in communication. The initiator of a conversation is responsible for ensuring accurate communication and uses feedback to test how well the message was understood. Just because the person to whom you have been speaking nods his or her head or makes a few comments does not mean he or she really got the message. Conversation often is open to multiple interpretations, depending on the listener's level of comprehension and the speaker's clarity.

As fleet managers, we regularly find ourselves trying to make the complex understandable to team members. Our job is to simplify the complex, sequence and prioritize the busy, and learn the technical. To check comprehension levels, fleet managers can ask different team members the same question several times. If fleet managers receive the same answer each time, they can feel reasonably comfortable they are being understood.

With lines of communication open, we can initiate actions in keeping with the strategic and tactical planning of our company. A simple example of this would be a strategic plan calling for reducing the fleet by 10% for the next fiscal year. We would begin by gathering the information needed to examine the 20% of vehicles with the lowest use rates, based on mileage over the past 12 months. We also would examine the 20% of vehicles with the lowest fuel use over the same time period.

With complete information, we can review usage, sorting out the necessary from the unnecessary and determining which vehicles should remain, which can be motor-pooled to increase usage, and which can safely be put in that 10% slated for removal—without jeopardizing productivity.

By gathering data on fuel use and mileage and by noting whether or not both sets of figures agree, we have the comparative real-life information needed to verify low usage.

Computers as Fleet Managers?

Here the computer enters the picture as a fleet management tool. An automated system promptly produces what is requested, without the error-prone human drudgery required to assemble the same information by digging through records or asking questions. On the other hand, it is possible to put too much faith in computers. Automated fleet management information systems do not actually manage fleets. These systems are information gatherers and arrangers that have been formatted by programmers to provide usable layouts.

Automated systems are neither smart nor threatening. They are simply another tool that fleet managers can use to reduce the time spent on information gathering and sorting. Thus, they also must be managed. Of the approximately 300 systems available today, at least one will match the needs and skills of any fleet manager who is looking to automation to help him or her do a better, more cost-effective, and energy-efficient job.

Choosing a system that is right for your fleet can be tricky. Be careful not to choose one that is so basic it soon is outgrown by the knowledge-seeking fleet manager who is using it. Functional managers drive programmers to improve present systems to meet the fleet manager's need to manage better and smarter.

In many cases, a fleet manager already knows—or "intuits" based on experience—the answer to a question. He or she uses the automated system simply to prove the answer.

For example, we know that the main costs to our fleet are tires, brakes, electrical, preventive maintenance, steering, suspensions, engines, cooling systems, air conditioning, and bodywork to repair damage. The automated system can supplement what we already know by telling us which specific components are costing the most within each category or application.

Let us use brakes as an oversimplified example. Light-duty vehicles experience more front-axle brake wear than rear axle because of weight shift and weight distribution per vehicle design. Heavier vehicles with vocational bodies (e.g., aerial buckets or dump trucks) have the most weight distributed to the rear axle, causing rear brakes to wear out quickly. Most fleet managers know this from doing brake work or supervising it, but the management information system will assign costs, validating the manager's hypothesis or showing productivity problems.

Simplifying the Complex

Part of the fleet manager's job is to help users and customers understand what is occurring with the vehicles they use by reducing complex systems to a simple explanation. An automated fleet management system helps support this process.

Numbers are objective. They are easy to understand and are not subject to much interpretation. The savvy fleet manager interprets the information provided by the numbers and combines it with his or her own knowledge to define a problem, identify the root cause, initiate corrective action, and cost it out, using real numbers from accounts payable and general ledger categories such as payroll, overhead, heat, lights, electricity, phones, and radios.

Management information systems for fleet maintenance have matured during the past 10 years, just as the fleet management profession has matured. With competition from outsourcers forcing fleet managers to keep an eagle eye on the bottom line, most fleet managers now pay attention to fleet costs and, to clarify in-house budget numbers, are taking an interest in cleansing data input, gathering meaningful historical information, and testing maintenance information to compare with and answer in-house budget number indications.

With historical data available, we can see clearly what we are really doing, and we can become more knowledgeable about the impact of what we are doing on our total budget.

Choosing an Information System

A management information system must pay attention to the accurate recording of maintenance-activity details and accurate input into the computer. This on-going process consumes 30% of your daily administrative resources.

The company budget has line items such as labor, parts, and fuel. If we have 10 mechanics working 80 hours on one shift, the question is: how much time is spent on what? The budget format verifies only 80 payroll hours. It does not identify who did what during those hours. Preventive maintenance requires 10 hours of labor on lights and 5 hours on suspensions. Another 5 hours are unaccounted. With a vehicle maintenance management information system, we can obtain more specific information on how those 80 hours were spent.

Here is an example of the output of a management information system on 80 hours made up of 40 scheduled hours and 40 unscheduled hours:

30 hours	Brakes (37%)—20 hours Scheduled (25%)
	10 hours Unscheduled (12%)
10 hours	Tires—Unscheduled (12%)
10 hours	Lights—Unscheduled (12%)
10 hours	Preventive maintenance—Scheduled (12%)
10 hours	Miscellaneous preventive-maintenance-generated repairs—Scheduled (12%)
5 hours	Suspensions—Unscheduled (6%)
5 hours	Unknown—Unscheduled (6%)

We now have some information about what we are doing and the opportunity to learn even more by following up on those 5 "unknown" hours.

Now, let us go through component control, using historical information generated by the computer.

We have coded our work orders with more than 200 repair codes associated with labor on work orders. Let us say that we have a total of 50 vehicles (birth to death) that are telescopic aerial devices (i.e., trouble trucks) with six years of life each.

The total lifetime cost for all 50 vehicles (Figure 6-12) is $759,548, with total parts at $363,978. That is 48% of the total cost, with labor making up the other 52%.

Repair	Type Desc.	Num. of Vehicles	Num. of R/O	Total	Comm.	Parts	Labor
117	Tires	50	659	$110,803	$3,021	$94,875	$12,906
113	Brakes, Replace, PM	41	162	57,943	33	31,324*	20,585
213	Brakes, Repair, Non-PM, Unscheduled	50	563	40,903	—	15,907	24,997
145	Engine Power, Replace, PM	28	43	39,679	—	26,968	12,693
245	Engine Power, Repair, Non-PM, Unscheduled	49	336	30,114	42	10,065	20,007
PM	Preventive Maintenance	50	1,501	68,124	—	3,410	64,714
116	Suspension	44	302	46,865	265	22,978	23,621
134	Lighting	53	1,616	40,129	—	11,671	28,458
115	Steering	44	319	32,273	106	14,326	17,842
	TOP 7 COMPONENTS			$457,347		$231,524	$225,823
						48.19%	51.81%
	ALL COSTS TOTAL			$759,548		$363,978	$391,282
* Clerical error; should be $137,324.							

Figure 6-12
Maintenance Cost of a 50-Vehicle Fleet

The 1:1 ratio of parts to labor is interesting. What is of further interest is that seven of the areas of repair make up 60% of the total lifetime costs and also are in a 1:1 ratio of parts to labor.

Digging further, we see that tires are the top cost, with $110,803 for 14% of the total cost. These are parts intensive, which makes sense because it takes 15 to 30 minutes to mount and install a $300 tire.

Further breakdown of the tire costs (Figure 6-13) shows that $25,000 of the total tire costs are tracks. That is a miscode that distorts the #1 cost total. Auditing the information is of utmost importance and is where your own experience, instinct, and intuition come into play to validate the data. You have the right and duty to be suspicious when you see items such as, in this case, tracks that are uncommon to the total.

Brakes:	
Front Brakes	$ 20,000
Rear Brakes	$ 10,000
Adjustments	$ 35,000
Compressor	$ 10,000
Front Drum	$ 15,000
Rear Drum	$ 10,000
Total	$100,000
Tires:	
Front	$ 35,000
Rear	$ 30,000
Trailer	$ 10,000
Rim Front	$ 5,000
Rim Rear	$ 5,000
Tracks	$ 25,000
Total	$110,000

Figure 6-13
Breakdown of Costs for Tires and Brakes

The next cost to consider is brakes, including scheduled brake replacement at preventive maintenance with predictive intervals and unscheduled brake repairs.

Brake replacement (scheduled) is a parts-intensive activity, whereas brake repair (unscheduled) is labor intensive. This makes sense because unscheduled activity tends to be labor intensive. The more efficient the preventive maintenance, the less unscheduled time will be required. What is disturbing here is that brake replacement parts and labor is $7,000 less than the total dollar number, which appears to be a human or software error. At $97,000 for brakes, this brings us to the #1 cost component.

Looking at Figure 6-13, which breaks down the $97,000 brake costs into selective component costs, front wheels and drums show a larger total expense than the expense for rear wheels. That does not make sense because in larger vehicles, the rear brakes do more of the stopping and thus should require more attention. Additional questions are necessary to validate this cost.

The aim of all these analyses is to pinpoint the most costly areas in order to focus on them to achieve cost reductions. The larger the expense, the more potential it offers for significant savings. By comparing information, we add to our store of information about this class of vehicle. Such information will help to improve our bottom line by pinpointing likely areas for cost savings.

In the case of brakes, areas of improvement can include specifying better brakes and slack adjusters, or procuring more application-specific foundation brake linings. Other things to consider would be using wider brake shoes and better brake drums, tires, lights, steering, and suspensions that may offer more "bang for our big buck" expenses.

The environment, condition of roads, work demands, and driving activity all impact cost. If we can focus on these factors, then we can be proactive in further reduction of costs through component life extension.

Figure 6-14 shows increasing costs for Years 1, 2, and 3, then a leveling of costs and minor reductions in costs for Years 4, 5, and 6. Years 1 to 3 are warranty years, and our consumable replacements (e.g., tires, brakes, filters) could be lowered through better specification of increased capacity or quality items to extend component life while reducing costs.

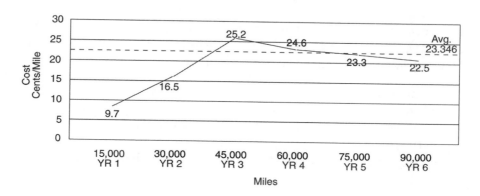

Figure 6-14
Maintenance Cost per Annual 15,000-Mi Interval

An example would be brake temperature. Off-the-assembly-line chassis brake temperatures may be targeted for 200°F. However, in real life, these temperatures might be more than 300°F. If we matched the correct foundation brake shoe and drum to the real-world temperatures, we would become more efficient. If we use this vehicle in three different application-specific environments (inner city, suburb, and city), then these specific domiciles and application areas would require appropriate components to be durable. Thus, we would specify the highest application-specific temperature brake shoes rather than the best brakes.

Years 1 and 2 are warranty-specific years, and any costs should be brought to the manufacturer's attention for resolution. A target dollar amount for warranty reclamation on which we should focus is 5% of the purchase price of a new vehicle. Thus, $75,000 per truck, multiplied by 50 trucks, equals $3,750,000, multiplied by 5%. This targets us at $187,500 in Years 1 and 2, which is 25% of the total cost we have recorded to look for a return from the manufacturer. Although we will be reimbursed for parts, we also should consider labor and downtime recovery. The best for which we can hope is the manufacturer or distributor doing warranty work on our property, keeping shuttle and downtime costs low and using their labor rather than ours.

When examining the ratio of costs for Years 1 and 2 (Figure 6-14), the Year 1 total usually is greater, which shows that warranty reclamation activity is documented and taking place. Because the cost in Year 2 is greater than the cost in Year 1 in this case (Figure 6-14), we should be suspicious that we are not paying attention to Year 1 warranty cost and reclamation. Although we may not recover the past warranty dollars in this case, we can start now for the present warranty dollar recovery. Through historical component costs, we will focus on areas to examine today for warranty recovery.

Managing budgeted dollars through fleet management information systems helps us to do real tracking in a proactive rather than reactive manner. A good offense (being proactive) is better than a good defense (being reactive). This also gets us on the scoreboards, keeps us competitive, and helps us cooperate with our customers by providing good service so that they bring in revenue with more of a profit.

Summary

In the preparations for the vehicle specification process, you should cost-quantify your areas of concern to analyze how you can lower your overall vehicle costs and extend vehicle life due to improved engineering.

Enforcement and education are important to reduce costs, but engineering is your most predictable control point.

The Purchasing Process

- Prepare a functional specification, rather than technical specification
- Buy what you need, not what you want
- Minimize modifications after delivery
- P.M. on time, with reasonable frequency
- Address driver write-ups in a timely manner
- Mechanic and driver training is a desirable support event
- Maintain good records and a record-keeping system
- Define the sales and award procedure

Chapter 7

Car and Light Vehicle Specifications

╬

Overview and Introduction

The car and light vehicle procurement process requires preparation of a specification describing the components desired and is based on the manufacturer's offering (Figure 7-1).

You can use the many publications available from commercial sources, such as the Kelley Blue Book, AIS Book, and PC Carbook available in hard copy or diskette to obtain the new-vehicle information.

These publications identify dealer invoice and suggested retail prices of the make, model (Figure 7-2), and its accessories by line items. A vehicle specification would be put together to establish a price for comparison to your budgeted forecast and to use as a reference for your bid evaluation comparison sent by bidders. You research current incentives offered by manufacturers and dealers and deduct them from your estimates.

The manufacturer charges the dealer for each vehicle the dealer receives and each vehicle the dealer orders. This price varies, based on the volume of sales achieved by the dealer. During one month, the dealer would turn over more or fewer vehicles, with the resultant price reflecting this volume. This price is referred to as the "tissue" price and is rarely made public.

YEAR:	
MAKE:	
MODEL:	
STYLE:	

STANDARD EQUIPMENT

ALL STANDARDS ARE 2002
2.9l (187 CID) MFI V6 ENGINE
3-SPD AUTOMATIC TRANSMISSION
525-AMP BATTERY
12-VOLT HIGH ENERGY IGNITION
100 AMP ALTERNATOR
16-GALLON FUEL TANK
RACK & PINION PWR STEERING
PWR 4-WHEEL DISC BRAKES
FWD
FULL SPORT SUSPENSION
P195/70R15 SBR BSW TOURING TIRES
5.5" STEEL WHEELS
15" WHEEL COVERS
DURABLE BASE-COAT/CLEAR-COAT PAINT
CHARCOAL WHEELHOUSE RUBSTRIPS
CHARCOAL BODY-SIDE MLDGS
FLUSH-MOUNTED TINTED GLASS
COMPOSITE HALOGEN HEADLAMPS

Figure 7-1
Format of Car Specification

New Car Discount Rate

$30,000	Price of car	Suggested retail price
4,500	15%	Retail mark-up 15%
$25,500		Factory dealer invoice price
600		Profit 2%
900		Fleet discount flat price 3%
2,100		Advertising 7%
$21,900		Targeted net price – 27% off retail price
		"Tissue" price

DESCRIPTION	INVOICE	SUG RETL
4 DR SDN	$11,888.69	$13,776.00
EQUIP GRP: COLOR-KEY F/R, FLR MATS,		
CRUISE CONTRL, TILT STRG WHL,		
GAGE PKG W/TACH, AM/FM STEREO,		
DIG CLOCK, DUAL R/C SPRT MIRRORS,		
PWR WNDWS, TRUNK OPENER, DR LKS	957.95	1,127.00
ELECT R WINDOW DEFOGGER	123.25	145.00
CLOTH BUCKET SEATS		
RECLINING SEATBACKS/ADJ HEAD RESTRAINTS	254.15	299.00
REAR SPOILER DELETE	- 108.80	-128.00
DESTINATION CHARGE	455.00	455.00
Total Model, Factory Options & Destination	13,570.24	15,674.00
+ Dealer Mark-Up	0.00	
+ Total Dealer Options	0.00	
Sale Total	13,570.24	
- Trade-in Allowance	650.00	
Trade Difference	12,920.24	
+ Trade Balance Owed	0.00	
+ Fees/Taxes/Adjust	0.00	
- Down Payment	0.00	
Total to Finance over 50 Months at 10.00%	12,920.24	
Monthly Payments of 317.03		

Figure 7-2
Factory Options

You can estimate the new-car discount rate at generally 25–30% of the suggested retail price, depending on the dealer vehicle inventory turnover and resultant volume discounts by the manufacturer to the dealer. This is reflected in the estimated dealer retail price.

- A $30,000 car costs a net of $21,900.
- The retail mark-up is 15% of the retail price.
- The profit is 2% of the retail price.
- The fleet discount is 3% of the retail price.
- The advertising is 7% of the retail price.
- Total targeted purchase cost is 27% of retail price.

This estimate varies, based on sales expectations and the economy. The dealer's price is further reduced by incentives based on volume of sales, spot rebates, and financing benefits, with the resultant price being the "tissue" price.

Dealers are required to stock a supply of vehicles made by the manufacturer and estimated by the manufacturer to be sold by the dealer. This is referred to as the floor plan. This is in addition to individual orders taken by the dealer.

Fleets are encouraged to place their orders early in the model year, for this volume is desired to provide the factory with work to supplement and maintain its initial assembly-line volume. Timing of this order by the fleets is critical for replacements. Depreciation starts at the beginning of the new model year. If a fleet orders in July or August, taking delivery in September or October, it can sell its vehicles at the end of their service life, taking advantage of their early model delivery in the sale price of the used vehicle. If a fleet orders a vehicle late in the model year, deliveries are at the end of the model year because they are fit in the assembly line where space is open from retail orders. The vehicle is one year old as soon as it is put in service. This results in a lower sales price, regardless of vehicle mileage at the end of its service life.

Of course, a fleet should not purchase a vehicle from dealer stock. The factory must make a special adjustment in its pricing to make it similar to a fleet price, which is not probable. Floor plan units will carry a higher price to fleet buyers due to the nature of the floor plan.

Even "perk" cars should not be purchased from the dealer's vehicle stock because these vehicles will carry a premium price. Order these vehicles at one time with the annual fleet purchase.

In putting together the vehicle specification, you should try to spec into the new vehicles the components that reduce maintenance cost. Typical maintenance costs are tires, brakes, and suspensions.

Because of better engineering, newer light passenger vehicles offer declining maintenance costs over 50,000 mi because of better ignition systems, antilock brake systems, cooling systems, starters, alternators, exhaust systems, EPA emission systems, and fuel injection systems. Pickups and vans follow suit.

Because of reduced life maintenance costs, you can consider extending non-perk vehicles to 100,000 mi or equivalent hours (40 mi = 1 hour), resulting in reduced light fleet costs.

Life-Cycle Cost Overview

Light vehicles are sensitive in life-cycle analysis to:

- Principal and interest costs (new vehicle versus depreciated old vehicle)
- Maintenance costs (new vehicle versus old vehicle)
- Resale value of old vehicle; new vehicle incentives; rebates; discounts

When the cost of the old vehicle is less to own and operate than the new vehicle, you should keep it another model year rather than replace it. The price of the new vehicle should include rebates due, and the resale value of the old unit should be subtracted.

Figure 7-3 accounts for the cost history of a seven-year-old vehicle at an estimated 13,000 mi per year. Figure 7-4 is a prioritized and sequenced

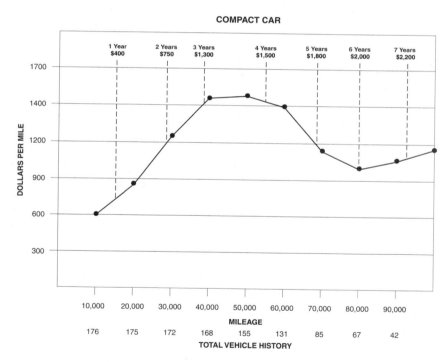

Figure 7-3
Compact Car Maintenance History

PM	Tires	Tires	Tires	Tires ....
Tires	PM	Brakes	PM	Brakes ....
Lights	Ign	PM	Brakes	PM ....
Ign	Brakes	Ign	Fuel	Suspension ......
Fuel	Power	Fuel	Exh	Power ....
Brakes	Fuel	Exh	Ign	Fuel ....
Power	Trans	Power	Power	Ign ....

COMPACT CAR

Present Car cost $310.50/mo X 50 Mos = $15,525

New Car Replacement Cost $438.75/mo X 50 mos (10% total increase for 3 years) = $21,937.50

Average Mileage - 13,000 Per Year - Compact Car

Resale Value History - $13,500 new cost

Time	Adjusted Value	Depreciation Rate
Resale Value Year 1	$9,660.00	(30%)
Resale Value Year 2	$7,728.00	(20%)
Resale Value Year 3	$6,182.40	(20%)
Resale Value Year 4	$4,945.92	(20%)
Resale Value Year 5	$3,956.74	(20%)
Resale Value Year 6	$3,165.39	(20%)
Resale Value Year 7	$2,532.31	(20%)
Resale Value Year 8	$2,025.85	(20%)

ESTIMATED MAINTENANCE COST/YEAR

Year 1	$ 400	-	13,000 miles
Year 2	$ 750	-	26,000 miles
Year 3	$ 1,300	-	39,000 miles
Year 4	$ 1,500	-	52,000 miles
Year 5	$ 1,800	-	65,000 miles
Year 6	$ 2,000	-	78,000 miles
Year 7	$ 2,200	-	91,000 miles

ACTUAL MAINTENANCE COST/13,000 MILES

0	—	13,000	$ 400
13,000	—	26,000	$ 750
26,000	—	39,000	$ 1,300
39,000	—	52,000	$ 1,500
52,000	—	65,000	$ 1,800
65,000	—	78,000	$ 2,000
78,000	—	91,000	$ 2,200

Figure 7-4
Repair Code Cost Summary

summary of the top seven repair costs that make up the bulk of the maintenance costs in each 20,000-mi segment.

Old Vehicle Versus New Vehicle Cost Comparison

The following examines the evaluation process.

At the end of Year 3, you should calculate the cost of the old vehicle to own and operate versus the new vehicle and project a Year 4 cost comparison. This analysis is dependent on resale values which vary depending on the market values.

Year 4 Cost of Old Versus Cost of New Vehicle:

	Old			New
Principal	$ 3,726	Principal		$ 5,265
Maintenance	$ 1,500	Maintenance		$ 1,215
Total	$ 5,226	Total		$ 6,480
Loss of Resale	$ 4,945	Resale Gain		−$ 4,945
Total Old	$10,171	Total New		$ 1,535

Net New Advantage $8,636 @ 52,000 mi

Year 5 Cost of Old Versus Cost of New Vehicle:

	Old			New
Principal	$ 621	Principal		$ 5,265
Maintenance	$ 1,800	Maintenance		$ 1,215
Total	$ 2,421	Total		$ 6,480
Loss of Resale	$ 3,956	Resale Gain		−$ 3,956
Total Old	$ 6,377	Total New		$ 2,524

Net New Advantage $3,853 @ 65,000 mi

Year 6 Cost of Old Versus Cost of New Vehicle:

	Old		**New**
Principal	$ 0	Principal	$ 5,265
Maintenance	$ 2,000	Maintenance	$ 1,215
Total	$ 2,000	Total	$ 6,480
Loss of Resale	$ 3,165	Resale Gain	– $ 3,165
Total Old	$ 5,165	Total New	$ 3,315

Net New Advantage $1,850 @ 78,000 mi

Year 7 Cost of Old Versus Cost of New Vehicle:

	Old		**New**
Principal	$ 0	Principal	$ 5,265
Maintenance	$ 2,200	Maintenance	$ 1,215
Total	$ 2,200	Total	$ 6,480
Loss of Resale	$ 2,532	Resale Gain	$ 2,532
Total Old	$ 4,732	Total New	$ 3,948

Net New Advantage $784 @ 91,000 mi

Life-Cycle Review and Analysis

At this time, 91,000 mi, major engine, transmission, and front-wheel-drive front-end maintenance risks increase, making replacement more desirable here. At the end of Year 6 and at 78,000 mi, the new vehicle is budgeted and ordered to replace the old vehicle between 91,000 and 100,000 mi. At this point, you would sell the 100,000-mi vehicle for an estimated $1,000 to $1,500. Whatever loss is sustained in resale is deducted from the net savings.

Higher annual mileage impacts this analysis. If 25,000 mi per year were put on a car, the resale value would be reduced significantly. If the vehicle is sold at 100,000 mi, it incurs higher annual maintenance costs (accumulated 2 years at 13,000 mi per year in one year's time) and does not benefit from reduced principal and interest costs because the vehicle still costs the full principal and interest in Years 3 and 4. High annual utilization requires a different strategy. Sell the unit at 50,000 mi, take advantage of the high resale value, and replace the unit every two years. Again, high annual utilization requires a different replacement cycle than low utilization.

Fleet image should be considered. If you desire to replace the vehicle at 30,000 to 40,000 mi and you choose not to keep it seven years because your product image is put at risk (e.g., a data process sales vehicle being old suggests old company data process technology, and a food sales vehicle being old suggests old company food stuffs), then this added cost is attributed to marketing and sales budgets.

Figure 7-5 shows a maintenance history for a compact pickup, with maintenance costs per 12,000-mi annual intervals. Figure 7-6 shows the top seven prioritized and sequenced repair costs for the 12,000-mi annual intervals for the compact pickup.

Figure 7-7 shows full-sized pickup maintenance costs per 13,000-mi intervals. Figure 7-8 shows the top seven prioritized and sequenced repair costs.

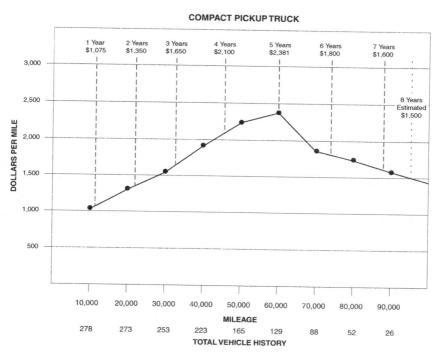

Figure 7-5
Compact Pickup Maintenance History

0-12,000	12-24,000	24-36,000	36-48,000	48-60,000	60-85,000
PM..............	Tires............	Tires...........	Tires............	Tires...........	Tires............
Tires...........	PM..............	Power.........	Power.........	Fuel............	Power.........
Lights.........	Power.........	Brakes........	Brakes........	Trans..........	PM..............
Fuel............	Fuel.............	PM..............	Ign...............	Brakes........	Brakes........
Power.........	Brakes........	Ex...............	Fuel.............	PM..............	Body..........
Cool...........	Lights..........	Ign..............	Lights..........	Lights.........	Fuel.............
Cab Door....	Ign...............	Lights.........	Susp...........	Power.........	Susp...........

Figure 7-6
4 × 2 Compact Pickups

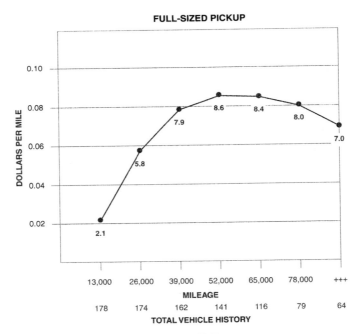

Figure 7-7
Full-Sized Pickup Maintenance History

0-13,000	13-26,000	26-39,000	39-52,000	52-65,000	65-78,000	78 ++
Tires	Tires	Tires	Tires	Tires	Tires	Tires ....
PM	PM	PM	Brakes	Brakes	Power	Brakes .....
Fuel	Brakes	Brakes	Cool	PM......... ...	Brakes	Power ......
Light.........	Ign	Power	Power	Trans	Ign	Cool ....
Cab Door .	Cab Door .	Trans	Fuel	Cool	PM...........	PM ....
Trans	Power	Ign	Ign	Ign.........	Cool	Trans ......
Wheel	Trans	Cab Door .	Cab Door .	Power	Trans	Ign ....

Figure 7-8
Full-Sized 4 × 2 Pickups

Due to better engineering pickup, among other things, light vehicles follow suit with the cars showing reduced maintenance costs. This could put these units in the same extended replacement cycle as the cars.

Note that the new-car industry sales fluctuate unpredictably, based on the economic environment. New light vehicle prices could decrease to stimulate sales. This, in turn, could decrease resale values of old vehicles, moving your seven-year benefits to lower ranges. With lower new prices due to rebates, it could motivate earlier replacements of cars and light vehicles, responding to economic recoveries in our country.

Also, if new light vehicle prices rise, used prices rise, too, stimulating old vehicle sales and further impacting your replacement strategy.

Light vehicle replacement analysis should be done annually to fit in not only with the economy of the country, but also operating strategy of your company. You could be growing and need additional vehicles, staying the same size with reduced or increased utilization of your fleet, or reducing your size, resulting in excessing some of your fleet, pushing off replacements.

The U.S. government has established Corporate Average Fuel Economy (CAFE) vehicle standards for manufacturers to meet. This impacts your operating fuel costs, which will be good cause to buy new for fuel savings.

Specification Construction

Light vehicle specifications need drawings when modifications are required to quantify product installations (Figure 7-9 through 7-11), noting sizes and quality of material which will be attached to aftermarket supplier specifications and vehicle manufacturer specifications.

When bids are sent out for response, bidders may quote items that you do not want. Their prices should be adjusted accordingly. For example, if a full-sized spare is required and the quote does not have it as extra, this price must be adjusted to be competitive, because the specification requires it. If bench seats

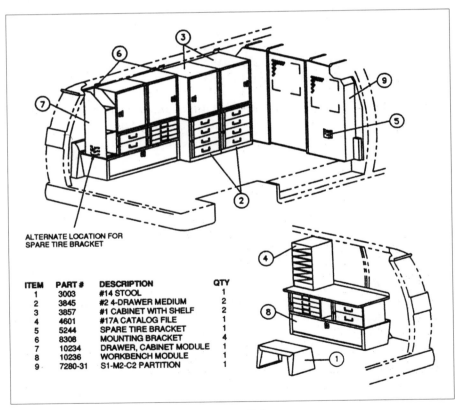

ALTERNATE LOCATION FOR
SPARE TIRE BRACKET

ITEM	PART #	DESCRIPTION	QTY
1	3003	#14 STOOL	1
2	3845	#2 4-DRAWER MEDIUM	2
3	3857	#1 CABINET WITH SHELF	2
4	4601	#17A CATALOG FILE	1
5	5244	SPARE TIRE BRACKET	1
6	8308	MOUNTING BRACKET	4
7	10234	DRAWER, CABINET MODULE	1
8	10236	WORKBENCH MODULE	1
9	7280-31	S1-M2-C2 PARTITION	1

Figure 7-9
Cutaway View of Van Cabinet Installation
(Courtesy Adrian Steel Co., Adrian, MI)

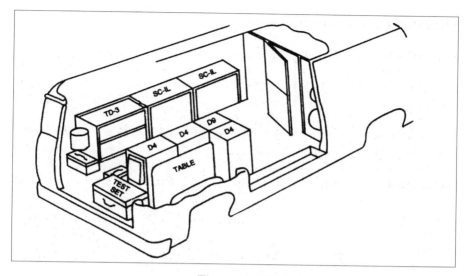

Figure 7-10
Side View of Van Cabinet Installation
(Courtesy Industrial Truck Body, Elizabeth, NJ)

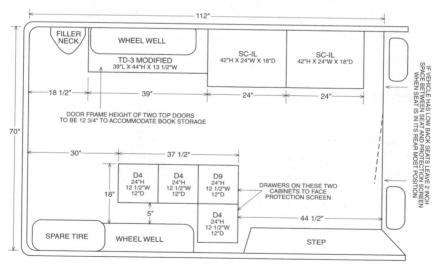

Figure 7-11
Top View of Van Cabinet Installation
(Courtesy Industrial Truck Body, Elizabeth, NJ)

are specified as standard and bucket seats are bid, it is another case for comparative price adjustments. Other areas include, but are not limited to, suspensions, warranty periods, and air bags.

The light vehicles should be shipped from the factory directly to your shops for in-servicing to capture delivery, preparations, and in-service savings completed by your own staffs, duplicating local dealers' activities that require supervision. Rather than incurring the supervisory costs of inspecting dealer prep costs, it is easier to do the in-servicing at your shops and know it is done with your own decals, marking, washing, phone or two-way radio installations, and other unique fleet tasks.

Thirty-day temporary license plates are required to give you time to register and insure your vehicles. When a new vehicle is put into service, remove the replaced vehicle at that time. If you do not do this, you will never have an opportunity to get the old vehicle out of service, and your fleet will increase in size unnecessarily.

Completing the in-service process of the new vehicle, the vehicle components (birth certificate) are input into the vehicle management information and reporting system to track its cost, schedule it for maintenance, and focus on warranty reclamation.

It is estimated that 2% to 5% of the purchase price of the car and light vehicle will be targeted for recovery in warranty from the manufacturer due to quality-related actions. (This can vary, depending on vehicle type and/or brand.) Careful assignment of these new light vehicles to an appropriate assignment will ensure "the loop is closed."

Note that from the time of ordering the new light vehicles, circumstances can change, shifting strategies of which the department should be aware because the budget reflects the agreed plans. Should these change, this should be brought up with those involved to ensure that the best interests of the company are served.

Chapter 8

Medium and Heavy Vehicle Specifications

╬

Today more than yesterday, you should spend time prior to putting together a medium- and heavy-duty specification to review with the user what he is doing today and what transportation can do to support his increase in productivity and to do it more efficiently with a better vehicle.

Many engineering changes have evolved to increase fuel economy and to provide a better driver environment, making overall efficiency improvements available in new vehicles.

Utilization

By carrying bigger loads, you will need proportionately fewer trucks. Therefore, fewer route miles are needed and less fuel is burned. This, combined with smaller trucks and their efficient routing, dispatching, and scheduling, can yield increased savings.

The table on the next page summarizes the possible savings that energy-efficient equipment and modifications can produce in your hypothetical truck with annual fuel costs of $20,000.

All these savings combine to make trucking 25–50% more fuel-efficient than it was in the pre-energy crisis days, provided trucks are properly operated at road conditions of 55 mph. It is important to realize that these percentage gains are incremental. Obviously, if you have a truck that is running with none of the modifications listed in the table, you can expect major improvements after one or two of the modifications are instituted. However, if a vehicle is fuel-efficient from the beginning, you will spend more to obtain increasingly smaller gains in miles per gallon. Eventually, you will arrive at a point where it becomes impractical to make additional investments in vehicle fuel efficiency.

Area of modification	Percent savings	Dollars saved per vehicles per year*
Aerodynamics (air foils, streamlining, etc.)	4–8	800–1,600
Rolling resistance (radial tires)	6–10	600–2,000
Power train (variable drive fans, etc.)	2–10	400–2,000
Speed control (governors, tachographs, etc.)	4–20	800–4,000
Vehicle maintenance	1–3	200–600
Driver training (shifting, etc.)	1–10	200–2,000
Operational techniques (better loading, routing, etc.)	4–8	800–1,600
Total possible fuel dollars and percent savings per year	22–69%	$3,800–$13,800

*Based on a truck with fuel costs of $20,000 per year.

Specification Areas for Better Vehicle Performance

The following components increase truck productivity. Whether being considered for installation on old vehicles or as positive features of new vehicles, they can tip the scales in favor of either holding on to old equipment or making new purchases.

In 1975, a typical tractor for 1 million miles maintenance costs $150,000/$165,000 versus a 1995 tractor maintenance at $65,000, per the following list.

Tractor Cost Differences Due to Technology
(Courtesy of Ryder Transportation Services, Miami, FL)

1975 $150,000/$165,000	(1,000,000 mi)	1995 $65,000
3–4	Engine Rebuilds	0
2	Transmission Rebuilds	0
2	Rear Axle Rebuilds	0
4–5	Clutch Replacements	1
8	Brake Jobs	3
83	Oil Changes	41
10	Sets of Tires	4
4	mpg	8

Additional engineering developments are multiplexing, alternative fuels, electric brakes, traction assist, antilock brakes, aerodynamic devices, 42 V, and fuel cell.

Economy Diesel Engines. High-torque, low-rpm diesel engines provide about 10% better fuel economy than gasoline engines, plus decreased operating and maintenance costs. They also increase vehicle life but preserve their marketing value because they will be in demand when they are sold. It is possible to obtain good fuel mileage with big engines if they are driven properly. The key is keeping both engine and road speeds down to engineered levels.

Clutch or Viscous-Drive Fans. Virtually standard equipment on most heavy-duty vehicles, these fans save horsepower by drastically reducing fan operation. Studies show that the fan is needed for as little as 5% of the running time of the engine, even in hot weather. Only proven fan components should be used because a 2% savings is critical. Greater fuel efficiency can be offset by higher maintenance costs on the fan drive.

Radiator Shutters. These shutters work with the fan and thermostat to keep the coolant temperature of the engine in its optimum range. Although these are something more to be maintained, they offer longer engine life through coolant condition control.

Radial Tires. Radials decrease rolling resistance and boost fuel economy by about 5%, but they cost more and, in a few applications such as low-speed and inner-city applications, they will not wear long enough to pay off. Radials can also yield greater tread mileage and can accept more recaps. If used on a tractor, as they are recapped they should also go on the trailer for maximum savings. Radial tires tend to pay back best on highway applications.

Wide-Step Transmissions. Economy engines are usually paired with gearboxes of fewer speeds, 5-, 6-, and 7-speeds being the common carrier favorites and 9-, 10-, and 13-speed transmissions popular among many private and small fleets.

Other arguments for smaller transmissions include the facts that fewer gears mean that there is less to go wrong and fix, and fewer gear changes mean less wear and tear on clutches. Multiple speed transmissions such as 9-, 10-, and 13-speeds offer the alternative economy of smaller engines to be used, which cost less but need well-trained drivers to operate efficiently.

187

Air Deflectors. Used for years by major fleets, air deflectors are coming into wider use by small fleets and owner-drivers who once disparaged the untraditional looks of the foil. They can be mounted on cab roofs, on the front of truck or trailer bodies, and even underneath the vehicle. Their makers usually claim fuel savings of at least 5%, with claims of 10% and more commonly made.

The aerodynamic devices work best at the upper speed ranges. Thus, truckers who cruise at 55 mph and higher will gain more from the foils than those who operate at lower speeds, such as in inner-city applications.

Lightweight Components. Trimming the unloaded weight of the vehicle saves fuel because there is less mass to move, and a lighter chassis leaves more room for added payload. Hundreds of aluminum body and chassis parts can be substituted for heavier materials, but they cost more and sometimes do not hold up as well. It's a trade-off, requiring investigation and a sharp pencil because of possible hidden maintenance costs.

Conventional Cab Design. Air flows over the bodies of long- or short-nose conventionals better than over most flat-faced cab-overs, so there is less frontal resistance. Also, conventionals generally cost less to buy and maintain, ride better, and offer better crash protection than cab-overs. Because of operating needs and length restrictions, of course, not everyone can use them.

Low-Restriction Air Intake and Exhaust. Low-restriction intakes and exhausts add efficiency by making the breathing chore of the engine easier. Devices such as frontal air intake, air rams, larger mufflers, and dual exhaust stacks add some weight and cost, but they will pay for themselves in 2% to 5% fuel savings.

Pusher or Tag Axle. The pusher axle mounted in front of the drive axle or the tag axle mounted in back of the drive axle eliminates the second driving axle and interaxle differential-power divider on tandem rear axle tractors or trucks. The reduction in friction can allow a tag-equipped vehicle to deliver 0.5 mpg more than two driving axles can. Many operators still insist on twin screws for added traction, but a pusher or tag axle is a sensible alternative.

Fuel Heater-Water Separator. Combined or in two sections, these devices keep wax particles small enough to pass through the fuel filters and take out the water that causes rust, which does not lubricate injectors, and could freeze and block the fuel system. They are simple, require little maintenance, and with the worsening quality of fuel today, are worth considering over insulated fuel lines and costly fuel additives.

Fuel Additives. Fuel additives are often needed to keep cold fuel flowing before it reaches a heater in the vehicle. They include pour-point depressants, wax modifiers, and water dispersants. Usually, they must be added at the time fuel is pumped into the vehicle tanks. This allows the pumping surge to suspend the additives thoroughly throughout the fuel.

Air Dryer. Air dryers are needed to keep moisture out of increasingly complex air brake systems. They are also good if the vehicle has an air-operated range-type transmission or air starting.

Air Starting. Users report less maintenance, better cold-weather cranking performance, and, in some cases, reduced weight with air starting systems compared with electric cranking systems. However, air systems add as much as $1,000 to the cost of a new vehicle and remain an unknown quantity to many truckers.

Crankcase Heater. If vehicle-mounted and electrically operated, crankcase heaters require getting the vehicle close enough to a "hitching post" or outlet in which to plug in the heater. Fleets sometimes use stationary boilers, which hook up to the vehicle via quick connect-disconnect fittings and circulate heated coolant. Either system keeps coolant or lube oil warm so the engine can be turned over easily.

Automatic Transmission. Available only from Detroit Diesel Allison, the automatic transmission is expensive, complex, and heavy. However, it is reliable and relatively trouble-free, even with untrained drivers. Automatic transmissions vastly decrease driveline shock, which means less maintenance and downtime. In most cases, they deliver better fuel mileage than manual transmissions deliver because of proper shifting cycles. They are finding wide acceptance, even among over-the-road operators and in multiple-drive applications. There is an additional up-front cost of $8,000 to $10,000 per vehicle which must be amortized.

Midrange Diesels. Primarily thought of as a fuel economy measure, smaller diesels for midrange vehicles also require less maintenance and last longer than gasoline engines. They usually pay back their higher price in less than two years, depending on their use. Benefits in terms of miles per gallon are twice those of comparable gas engines.

Premium Paint. Polyurethane or urethane looks better and lasts longer than standard paints and costs only a couple of hundred dollars more. That difference should be made up in less work and expense in keeping the vehicle looking good, and in a higher trade-in value. Tasteful, multicolor paint schemes can also pay off at trade-in time.

Premium Electric Components. Color-coded harnesses can cut maintenance time, as can heavy-duty or sealed wiring or conduit. Shock-resistant light fixtures can greatly extend bulb life and reduce the labor time needed to change the bulbs. High-capacity, brushless alternators will reduce charging problems and maintenance time. Maintenance-free batteries will soon pay off their extra cost in saved labor expenses alone, if mechanics and/or technicians are trained to service them.

Gauges. A source of maintenance problems, gauges are nonetheless popular adornments on the dashes of most trucks. Some items cost little, look good, and make the driver happier and more productive. Gauges also can enhance the trade-in value of the vehicle.

Chrome Exhaust Stacks. This so-called ultimate extravagance actually lasts longer than the standard steel stacks and, if the owner is committed to keeping vehicles clean, makes that chore easier.

Upgraded Ladders and Handholds. Improved ladders and handholds make cab entry and egress easier and safer and will probably pay for themselves in reduced workers' compensation costs.

Premium Interior. A few extra dollars for fine upholstery and padding on doors and roof make a rig more attractive and comfortable. These could encourage the driver to take better care of the vehicle.

Cruise Control. Cruise control allows steady, more economical cruising and makes driving safer and less tedious. Estimated fuel savings on highway and interstate roads are 0.5 to 1 mpg increase.

Shock-Mounted Lights. Shock-mounted lights can be used to extend light bulb service.

All these features are designed to save time and money. Attention focused on choosing the right specs can be wasted, however, if the driver does not know how to obtain the best performance from the vehicle. Drivers should be instructed on how the vehicle should be operated and given guidelines on things such as road speed and idling. Such rules should be enforced through supervision, to the extent of even monitoring with tachographs or electronic devices.

If you are buying trucks for a new operation, learn everything you can about the requirements. Where will the vehicle operate (city, suburb, interstate), and on what type of terrain (flat, hilly, mountainous, a mixture)? At what speeds will it operate in its environment? If you are adding to the number of, or replacing, existing trucks, dig into the weak points of the present fleet. Do not buy new units that will give you the same problems as the old ones.

Comparison of Gasoline and Diesel Engines

Gasoline Engines:

1. The initial cost is lower.

2. Fuel consumption is higher.

3. Gasoline engines are lighter than diesels.

4. Gasoline engines require more service attention than diesels; however, with developments in electronic ignition systems, there will be less maintenance, and spark plug life will be longer.

5. The intervals between overhauls are shorter with gasoline engines, but this depends on the quality of drivers and maintenance programs.

6. Replacement parts cost less.

7. Higher governed speeds allow a more flexible engine.

8. Less engine torque allows the use of lighter-weight components to transmit power to the rear wheels.

9. Gasoline engines have inefficient idling characteristics.

10. Parts are readily available.

11. Gasoline engines have good cold-weather starting characteristics.

12. The lower starting torque of gasoline engines requires less battery capacity.

13. Mechanics need less specialized knowledge.

14. Test equipment required for carburetors, fuel injection, and ignition troubleshooting is less complex and more expensive.

15. Fuel is more widely available.

16. There are problems with complying with emission control standards.

17. Fuel injection and upgraded electronic ignition and diagnostic systems reduce maintenance costs.

Diesel Engines:

1. There is high initial cost, running as much as $3,000 more than comparable gasoline engines.

2. Diesels have good fuel economy and can get up to 100% better mileage than gasoline engines.

3. Diesels are heavier, not only the engine block but the transmission, front suspension, axles, and drivelines as well. The higher engine torque requires these stronger components.

4. The combustion system is free of electrical maintenance, but diesel injectors require clean fuel, free of water.

5. Diesels can go farther between overhauls, provided drivers do not abuse the trucks and mechanics do a respectable PM job.

6. Replacement parts are expensive. Rebuilding diesels can cost up to three times what it would cost to rebuild a gasoline engine.

7. Lower governed speeds mean lost productivity if a truck cannot keep up with traffic and can lead to more engine lugging.

8. Heavier components mean that heavier frame rails, bracketing, and other parts are needed to support the heavier components.

9. Diesels have better idling characteristics than gasoline engines do and can idle four times as long on the same amount of fuel.

10. Parts availability can be a problem, especially on foreign engines.

11. Cold-weather starts for diesels require either assistance or block heaters.

12. Because of their high starting torque, diesels require either more battery capacity or air starters.

13. Mechanics require specialized knowledge to work on diesel injection systems.

14. There is greater opportunity to apply a diesel to the wrong sort of job or to use it incorrectly and be saddled with catastrophic engine failure.

15. Number 2 diesel fuel requires blending with Number 1 and/or additives to prevent clouding or jelling in winter.

16. Diesels require emission control systems, which are being met with turbocharging electronics and low-sulfur fuels.

17. Diesel scheduled maintenance is 50% less frequent than gasoline engines.

Figure 8-1 shows a comparative cost of maintenance of a gasoline versus diesel engine in a Class 7 vehicle, noting $919 more per year average cost over eight years above a naturally aspirated diesel engine. Most of this diesel savings becomes available through higher gasoline engine maintenance costs in Years 5 through 8.

Tire Selection

Proper tire selection is a basic weapon in the battle to keep truck operating costs at a minimum. There are several points to remember, and they must be considered simultaneously because tire problems are complex. You may have to compromise in one aspect of tire selection to get exactly what you need in another.

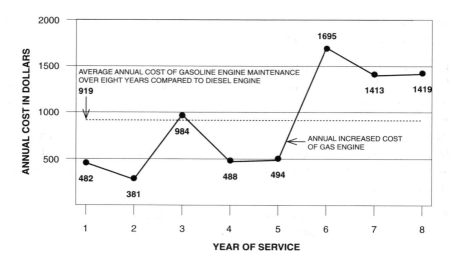

Figure 8-1
Analysis of Gas Versus Diesel Engine,
Chassis-Related Repair Codes Only

Consideration of vehicle type and the use for which it is intended are important. You must decide whether to accept original equipment tires or request the larger or smaller size options. However, options are limited by the vehicle clearances and recommended rim size. For instance, if a truck is intended for long-legged, over-the-road operation, tires should be chosen for mileage and safety.

What loads will you be carrying? Specify enough tire to support the maximum loads. You may not be loaded to the maximum each time, but when you are, you will need that stronger tire. It is never desirable to exceed the specified load range (or carrying capacity) of a tire. Shifting loads and increased loading cycles should also help determine tire size.

A major factor in tire selection is the speed and continuity of your operation. For instance, long, high-speed runs produce tire heat, and heat is one of the prime causes of tire failure and rapid wear. For long, high-speed hauls, you should use tires with cooler-running characteristics, such as radials. A low-speed tire would be a tube-type, bias-ply, or bias-belted tire. A low-speed tire should have strong sidewalls, puncture-resistant tread, and a tread design to offer maximum traction in its operating environs.

Consider the tread design that best suits your operation. Rib-type designs are best suited for front wheel positions. However, they may also be used on drive wheels of tractor-trailers and larger straight trucks. Lug types are best for drive wheels where traction is needed in high-torque service. On drive wheels, lug designs usually deliver more tread miles than do rib designs. They can also be moved back to trailers for additional service. The "severe" lug designs should be used only in a combination of on-off road applications, areas where traction is essential.

Buy quality new tires that have a good reputation for retreadability. By retreading, you obtain maximum life from your tire casings.

Tire Types: A Case for Radials

Tube-type tires are often the best choice for low-speed operations; however, tubeless tires offer these distinct advantages over tube-type construction:

1. **Simplicity.** The tube, flap, side ring, and locking ring are eliminated. Tubeless tires require only a one-piece wheel. They are safer and easier to mount.

2. **Weight savings.** Fewer component parts per assembly are required.

3. **Fewer road delays.** Penetrating objects such as nails are gripped more tightly, helping to prevent rapid air loss until a permanent repair is made.

4. **Labor savings.** The fewer parts per assembly and the fewer road delays result in labor savings.

5. **Cooler running.** The elimination of the tube and flap results in cooler running characteristics.

Radial tires will generally produce a 6–10% mpg gain, owing to their lower rolling resistance. They also last longer and give better traction. If you need to replace your tires, consider the advantage of a radial tire not only as a fuel-saving device but also as a cost saver because of its recapability and longer casing life. The radial tire is gaining increased popularity as trucking costs escalate.

Fuel Economy

Radial body construction gives less resistance to rolling than does that of bias-ply tires. Less rolling resistance means better fuel economy. Controlled proving ground tests show up to 6% fuel savings at steady highway speeds for radials, compared with conventional bias-ply tires. This can mean big savings when considered for many vehicles over may miles.

> Example: A fleet of 10 over-the-road tractors averages 90,000 mi annually. The average fuel usage with bias-ply tires is 1 gal every 5 mi, or 5 mpg.
>
> 10 tractors × 90,000 mi = 900,000 mi annually
> 900,000 mi / 5 mpg = 180,000 gal of fuel per year
> 180,000 gal × $1 per gal = $180,000 annual fuel cost
>
> If by switching to radials, this fleet experienced the same 6% fuel savings that engineers have measured in their controlled tests, the annual savings would be: $180,000 × 0.06 = $10,800, or $1,080 per tractor.

Reduced Downtime

The tough steel cord body and four additional tread plies of steel in radial tires form a protective "steel shield" against road hazards and impact breaks.

The specially blended tread and sidewall compounds, coupled with the steel shield, mean substantial reductions in tire-related downtime (up to 40% less compared with bias-construction tires, according to commercial fleet tests).

> Example: A fleet of 10 units, tractors and trailers, averages 5 flats per month. The cost of 1 flat (downtime, service charge, repairs) is $100. The annual cost is 5 flats × $100 per flat × 12 months = $6,000.
>
> If an actual fleet experienced the same 40% reduction in downtime observed in commercial fleet testing, it could expect savings of:
>
> $6,000 × 0.40 reduced cost = $2,400 savings per year, or $240 savings per unit.

Analyze the Job

The next step in purchasing new vehicles is to set up your specifications list. You may be able to write your own if you have qualified personnel to do it, or you can get help from knowledgeable sales engineers at dealers or factory branches.

Study the job requirements for your trucks. Know what the trucks will haul, how the cargo will be shipped (Figure 8-3), how much the cargo weighs, and the quantity per trip. Know about product handling. What are the dimensions? How many cubic feet, cubic yards, or gallons of carrying capacity are needed? A chassis should be selected that can handle your loads with sufficient capacity for the occasional load that exceeds the norm.

Loading and Unloading Methods

Loading characteristics influence component selection, especially horse-power requirements, gearing, and suspension systems. It is important to determine if there are full loads both ways or one way, partial loads, or diminishing loads.

Routes

Because you have checked the basics, you now must fit your data into the total operation. Careful specification at the start can save big money later, over the life of the vehicle. In what type of service will the truck be used? Where will it run? City streets? Open highways? Mileage and hours are also important. How many miles will the truck run per day, per week, or per round trip? How many stops will it make per day? Next comes the type of terrain, a consideration in determining horsepower needs and gear ratios. Check the maximum grade the truck must negotiate. Is there an inclined loading dock or some other steep incline in an unlikely place? Starting ability can be a worse problem than gradeability. This can be as important for local trucks that are used now and then on expressways as it is for long-distance vehicles. Units you may buy locally for use in another city or state must be specified for local terrain there— even if they are doing similar work.

Whether the truck will be involved in day versus night operation has a bearing on electrical and charging system needs, as does the number of lights. Note any unusual operating situations such as extreme temperature conditions, high altitudes, narrow streets, and low bridges.

Check Current Equipment Specifications Closely

If you are replacing vehicles, a close look at your old specs will be most revealing. Have you suffered repeated engine failures, frequent clutch replacement, broken suspensions, shortened tire life, frame failures, U-joint and driveline wear, accelerated brake lining wear, or electrical system complaints? If so, your present specs could be the root of the problem. Drivers could also be a contributing factor.

Do not buy new units that will give the same trouble again. Often, talks with drivers and mechanics will uncover problems and give you ideas on how to solve those problems. You must put their comments in the proper perspective, however.

If you plan to transfer an old body to a new chassis, do the old body specs and dimensions fit the new chassis wheelbase, and is the cab-to-axle (CA) dimension the same? If you are buying a new body, check the old ones first to find the trouble points so the new ones can be properly specified to reduce wear and tear.

Write down what you need, and review your findings with experienced manufacturing salespeople. You will end up with a standard request for a bid that each supplier can use as the basis for an accurate bid on identically equipped models. Writing vehicle specifications eliminates the possibility of oral misunderstandings.

To obtain the right truck for the right job, you must know what the job is. The right truck should be the least expensive over its life cycle.

Developing Truck Specs

There are three types of truck builders:

1. **Integrated**—They make all the components themselves. (An example in the United States is Mack Trucks.) All components are made and installed on their chassis.

2. **Assembled**—These are mixtures of manufacturers' engines, transmissions, and drivelines, some customized. (Mack will assemble also.)

3. **Pre-specified**—The manufacturer makes trucks as an off-the-shelf item, backed with an energetic warranty program to ensure efficiency.

Weight distribution, gradeability, and geared road speed are only a few of the factors affecting vehicle performance. A truck dealer's sales representative can select the right truck to meet a customer's operating requirements. To guide salespeople in their choices, vehicle manufacturers prepare graphs, charts, and formulas. These sales aids should be used to make an intelligent equipment choice in combination with the fleet manager's specifications. Close comparison of the two resources should explore any glaring deficiencies before a faulty purchase is made.

Chassis cabs and axle locations vary and have different applications. Conventional cabs tend to ride better and have less air resistance. Turning radius is restricted, and care should be used in the correct work assignment.

Cab-over-engine chassis offer longer bodies to be mounted on the frame and a more efficient turning radius. They tend to ride harder for the driver because the driver's seat is over the front axle. Also, more air resistance is offered, increasing engine horsepower needed to overcome this air resistance.

The items shown in Figure 8-2 describe cab configurations available for your choice and proper application.

The fundamentals of truck selection are based primarily on two requirements: 1.) the ability to carry the load, and 2.) the ability to move the load. Analyzing these requirements takes several steps, which will be covered in the next few sections.

CAB TYPE

Low Cab Forward—Cab is positioned forward of front axle; single or tandem drive axle

Cab Over Engine (COE)—Cab is positioned directly over front axle; single or tandem drive axle; with or without sleeper

Short Conventional—Bumper-to-back-of-cab (BBC) dimension is less than 96" (do not include sleeper compartment); single or tandem drive axle

Medium Length Conventional—BBC dimension is 96" to 112" (do not include sleeper compartment); single or tandem drive axle

Long Conventional—BBC dimension is more than 112" (do not include sleeper compartment); single or tandem drive axle

Severe Service Conventional—Generally BBC is more than 112"; steel (diamond-plated) hood and fenders; double channel, reinforced frame

FRONT AXLE ARRANGEMENT

Axle Forward—Front axle is positioned as far forward as possible to increase the dimension between the front axle and the drive axle(s)

Axle Back—Front axle is set back from the bumper a minimum of 42" to transfer maximum weight forward and increase maneuverability

---

Figure 8-2
Definition of Terms

The weight distribution on the front and rear axles is proportional to the cab and chassis configuration. Conventionals offer 20–25% of the weight on the front axle, whereas cab-over configurations offer 25–33% of the weight on the front axle.

Carrying the Load

The first step is to carefully study the gross vehicle weight (GVW) and have it properly distributed (Figure 8-3). Correct distribution of weight to the axles provides the best ride, steering control, and maximum driving and braking traction and guards against premature failures due to overloaded springs, axles, and tires.

Example: Suppose you wanted a truck with a 16-ft body that was 96 in. wide and 84 in. high. Figure 8-5 is an example of a weight distribution chart you might develop.

You estimate that the body and expected payload weight will add up to 14,500 lb. Checking the manufacturer's data book for a unit that would

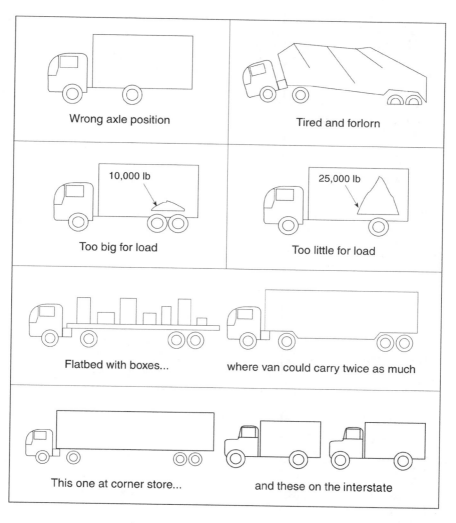

Figure 8-3
A Gallery of Equipment Use Faults

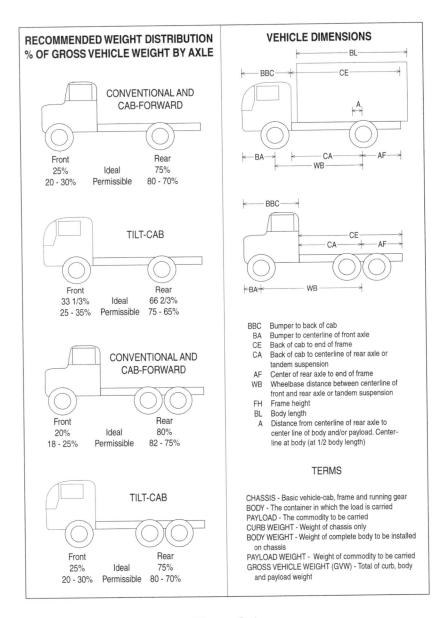

Figure 8-4
Recommended Weight Distribution Percentages

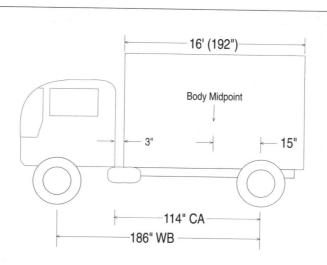

WEIGHT, POUNDS

Front Axle		Rear Axle	Total
3300	A. Chassis	2500	5800
200	B. Body	2300	2500
1000	C. Payload	11,000	12,000
4500	D. Total	15,800	20,300
22%	E. Percentage of GVW	78%	

COMPONENT CARRYING CAPACITIES, POUNDS

5000	F. Axles	16,000
5200	G. Springs	16,200
5480	H. Tires	15,800

Manufacturer's GVW rating: 20,600 pounds

Figure 8-5
Straight Truck Weight Distribution Chart

accommodate this weight, you find that the chassis weight should be about 5800 lb. The combined body, payload, and chassis weight is now approximately 20,300 lb. Manufacturer's specifications might then indicate that the chassis GVW rating must be at least 20,600 lb.

The next step is to choose a wheelbase with a cab-to-axle dimension (CA) suitable for the 16-ft body. Using a rule of thumb that says roughly 60% of the body length should be ahead of the rear axle centerline, a CA of 118 in. (including 3 in. between cab and body) is required. The closest you can come in the line of equipment off the shelf of a particular manufacturer being considered might be 114 in., which is available in a cab-over unit with a 186-in. wheelbase. (Each manufacturer has slightly different dimensions, and individual requirements may affect the wheelbase and CA choice.)

Looking at Figure 8-5, the chassis weight (A) of 5800 lb is distributed with 3300 lb on the front axle and 2500 lb on the rear, as would be indicated by the manufacturer's specifications. Of the 2500 lb of body weight (B), 200 lb is on the front and 2300 is on the rear. The 12,000-lb payload (C) is divided into roughly 1000 lb on the front and 11,000 lb on the rear. Note: Body and payload distribution are found by multiplying each weight by the distance that the centerline of the body or payload is ahead of the rear axle centerline, in this case 15 in., and dividing the answer by the wheelbase. (Body length is 192 in. total. One half of the length is 96 in. The cab to rear axle length [CA] is 114 in. CA minus midpoint, 114 – 96 = 18 in. Subtract the 3 in. between the back of the cab and the front of the body, 15 in. Payload weight = 12,000 lb × 15 in. = 180,000 ÷ the wheelbase 186 in. = 967 lb on the front axle. Body weight is 2500 lb × 15 in. = 37,500 ÷ the wheelbase 186 in. = 201 lb on the front axle.) This will give the weight in pounds on the front axle. This is subtracted from the total body or payload weight to obtain the portion distributed to the rear axle. As can be seen in Figure 8-5, the total axle loads (D) therefore come to 4500 lb on the front axle and 15,800 lb on the rear for a gross vehicle weight of 20,300 lb. The percentages of gross weight (E) carried by the front and rear axles would be 22% and 78%, respectively. This unit does not balance well. Thirty-three percent of the weight should be on the front axle and 66% on the rear axle.

To carry these weights, the components specified for the axles (F) would be a 5000-lb front axle and a 16,000-lb rear axle. They have slightly more capacity than required. However, these are standard components and are well-suited to the application. The vehicle will be light in the front, resulting in a rough ride for the driver.

The springs (G) are more than adequate. The two 2600-lb front springs are standard, for a total capacity of 5200 lb. The rear springs, rated at 8100 lb each, for a total capacity of 16,200 lb, are slightly heavier than required but provide stability as well as carrying capacity for occasional overloads and stressful driving conditions. This will lead to a bouncing front end.

The tires (H) are 8.25 × 20 twelve-ply rated. The single-tire load capacity for the front axle application is 2740 lb for a total of 5480 lb for two tires at 50 psi. On the rear, at 3950 lb for each of the four tires at 90 psi, the total capacity is 15,800 lb. (Note: Adequate tire capacity could have been achieved by splitting tire sizes and using 8.25 × 20 ten-ply on the front and 9.00 × 20 ten-ply on the rear. This creates a maintenance problem, however. Mixing tire sizes should be avoided.)

The unit will not balance well, and the load-bearing components are up to capacity. However, without a weight distribution chart such as Figure 8-5, how would you know? The sales representative will check the data. The vehicle needs more weight on the front axle, and it should be weighed before you accept delivery to ensure that the vehicle that was built for you is what was calculated on paper. Computer assisted design (CAD) will identify components and their assembled weight (Figure 8-6).

The bidders need to submit a weight distribution analysis with the bid. When the bid is awarded and the vehicle is delivered, the vehicle can actually be weighed prior to acceptance, verifying the final weights.

If the weights are not accurate, this problem can be brought up to the manufacturer for correction. If not corrected, the company does not have to accept this vehicle (Figure 8-7).

Dimension Calculation Formulas
from the National Truck Equipment Association

To select the proper body length to match a given chassis:

$$BL = \frac{(GAWR.R - CWR) \times WB}{GVWR - CW - WB + CA - CB} \times 2$$

DESCRIPTION	WEIGHT	CG	WEIGHT ON FA	WEIGHT ON RA
100 Gal Fuel	700.	145.14	518.	182.
1040-01 Saddle Pack	1200.	87.00	533.	667.
158" Subframe	1580.	28.75	232.	1348.
16" Flatbed Body	1900.	-14.00	-136.	2036.
Auxiliary Outrigger	1100.	111.75	627.	473.
Basket Access	165.	-50.43	-42.	207.
Basket Liners (2)	80.	-74.00	-30.	110.
Basket Rotators (2)	80.	-74.00	-30.	110.
Boom Rest	150.	110.41	84.	66.
Capstan Assembly	302.	-80.81	-125.	427.
Driver and Passengers	600.	148.16	454.	146.
Main Radial Outriggers	1400.	-58.06	-415.	1815.
Manual Jib and Winch	160.	-74.00	-60.	220.
Outrigger Pads and Storage	60.	.00	0.	60.
Payload	4500.	8.25	189.	4311.
Pintle Hook Assembly	100.	-91.12	-46.	146.
Rear Access Step	20.	-93.50	-10.	30.
Second Basket	130.	-74.00	-49.	179.
Side Access Step	45.	58.00	13.	32.
Sign Racks (2)	50.	122.50	31.	19.
T-5055 Aerial Tower	6750.	50.50	1739.	5011.
Water Cask Compartment	50.	112.82	29.	21.
Wheel Chocks and Storage	16.	45.21	4.	12.
Bare Chassis Weight on Front Axle	8485.		8485.	0.
Bare Chassis Weight on Rear Axle	7632.		0	7632
TOTALS	37255.		11995.	25260.

The Above Analysis Is Based on the Following Chassis Information:

FRONT AXLE WEIGHT	DESCRIPTION	REAR AXLE WEIGHT
6536	Ft. 900 6x4 (126" CT) Bare Chassis	5989
-26	6 Cyl. 240-HP Diesel Engine	0
316	Fuller T8607 7-Speed Transmission	72
1200	Marmon Herrington MT-17 Front Axle	880
0	Rockwell RT-40-145 Rear Axle	91
216	Double Channel Frame with 26.5 Section Modulus	506
38	Chopped Frame Extension (18")	20
175	Dual 50-Gal Fuel Tanks	54
10	130-AMP Motorola Alternator	0
20	Bendix AD4 Air Dryer	20
8485	TOTALS	7632

Figure 8-6
Weight Distribution Analysis
Aerial Device
6 × 6 Chassis 196" WB 126" CT
(Courtesy Baker Equipment, VA)

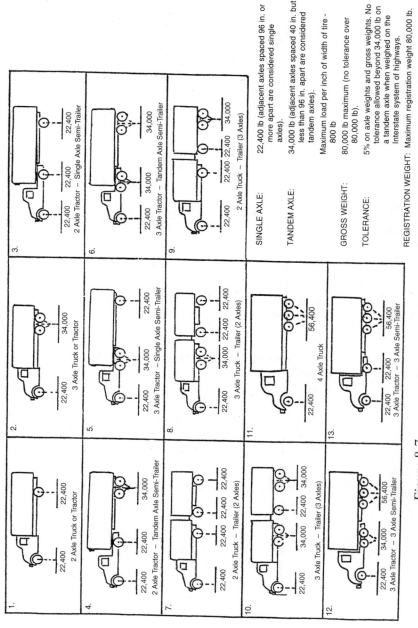

Figure 8-7

Axle Weight Limitations, Title 39:3-84

207

To select the proper wheelbase to match a given body:

$$WB = \frac{AC + CB + (BL/2)}{GAWR.R - CW} \times (GVWR - CW)$$

To select the proper cab to axle length for a given body:

$$CW = WB + CB + (BL/2) - \frac{WB \times Pr}{P}$$

where

WB	=	Chassis wheelbase
CA/CT	=	Cab to axle or cab to center of tandem
AC	=	Front axle to back of cab (BBC – BA = AC)
BA	=	Bumper to front axle
BBC	=	Bumper to back of cab
CB	=	Cab to body clearance
GVWR	=	Gross vehicle weight rating
GAWR.F	=	Gross axle weight rating–front
BAWR.R	=	Gross axle weight rating–rear
CW	=	Curb weight
CWR	=	Curb weight–rear
P	=	Total payload including body
Pr	=	Payload rear (GAWR.R – CWR = Pr)
BL	=	Body length

If the sum of the two GAWRs exceeds the manufacturer's GVWR, you will first need to reduce the GAWR in the formulas so that the sum of the axle ratings is equal to or less than the manufacturer's GVWR. This can be accomplished by dividing the GVWR by the sum of the GAWRs to obtain a percentage, and then reduce (multiply) the GAWR by the percentage. The result is a new GAWR that should be used in the calculations.

Vehicles with their weight not distributed properly offer driveability problems and premature failures amounting to unnecessary costs.

Tractor-Trailer Weight Distribution

As with the straight truck, proper weight distribution in the tractor-trailer combination is important. The only way to select a proper vehicle is to analyze the choice before purchase by constructing a weight distribution chart. If you can find a similar unit in use, drive it to see how it works.

Making a weight distribution analysis for a tractor-trailer combination is more complex than it is for the straight truck. It should be made twice: once for the trailer, and once for the tractor. The method is the same, but the terminology is different (Figure 8-8 and Figure 8-9).

Dimensions and Terms

The tractor A dimension is the distance the fifth wheel is in front of the center of the rear axle.

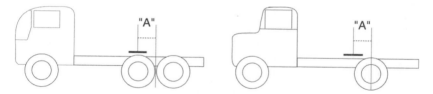

Payload Versus Axle Load

Weight reaching the highway is distributed as shown by the arrows. Load on front axle is determined by distance that kingpin is ahead of tandem.

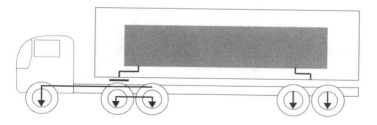

Figure 8-8
Tractor Trailer Weight Distribution

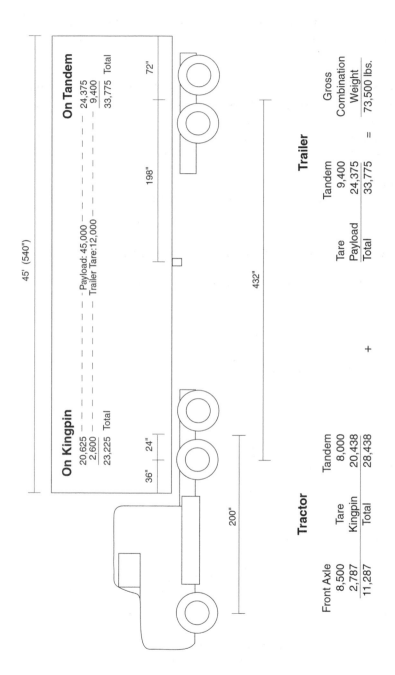

Figure 8-9
Tractor-Trailer Weight Distribution Chart

The trailer must be treated as a separate vehicle. Basically, it is nothing more than a large body with wheels under the rear and a kingpin up front for coupling to the tractor.

All equipment should be analyzed as to its ability to carry the load, including its vehicle wheelbase and component speedability and gradeability. If this is truly a new vehicle, a prototype may be considered. If the vehicle is new only to you, the salespeople can locate a similar unit, and you should inspect it. Try it under your conditions to verify its capabilities.

1. Weight of Trailer   12,000 lb
    (2600 lb Kingpin, 9400 lb Rear Tandem)

    Payload   "Water Level"   45,000 lb

    Trailer "A" Dimension—Center of Trailer to Center of Tandem 198"

    Wheelbase—Center of Trailer Kingpin to Center of Tandem

$$KPL = \frac{Payload \times A}{Wheelbase}$$

$$KPL = \frac{45,000 \ lb \times 198 \ in.}{432 \ in.}$$

$$KPL = 20,625 \ lb$$

Tandem Load   =   24,375 lb
Total Load Distribution   45,000 lb

Kingpin Weight   =   2600 lb + 20,625 lb   =   23,225

Tandem Weight   =   9400 lb + 24,375 lb   =   33,775

Total Loaded Weight   57,000

2. Tractor Front Axle   = $\dfrac{KPL \times A}{WB}$ = $\dfrac{23,225 \times 24 \ in.}{200 \ in.}$ (KP to ctr = 2787 lb of Tandem)

    Front Axle   =   2787 + Chassis 8500 = 11,287 lb

    Tractor Rear Axle   =   Payload + Wt of Chassis on Rear Wheels
                                23,225 + 8000 lb − 2787 = 28,438

Final Distribution of Weight:

F.A. 11,287 + 28,438 RA + 33,775 = 73,500 lb

The engine is the heart of the efficiency of any vehicle. Too low a horsepower rating, and premature failures and high repair costs occur. Too large an engine, and you will spend an excessive amount of money for an engine that provides poorer fuel economy and no extra benefits. Figures 8-10 and 8-11 show three areas of concern:

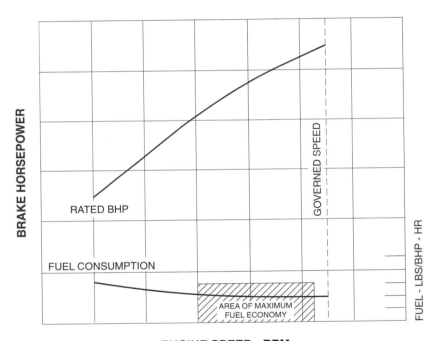

ENGINE SPEED - RPM

Figure 8-10
Gear Your Trucks to Take Advantage
of the Maximum Fuel Economy Area
(Courtesy Motor Truck Engineering Handbook,
James W. Fitch, Publisher)

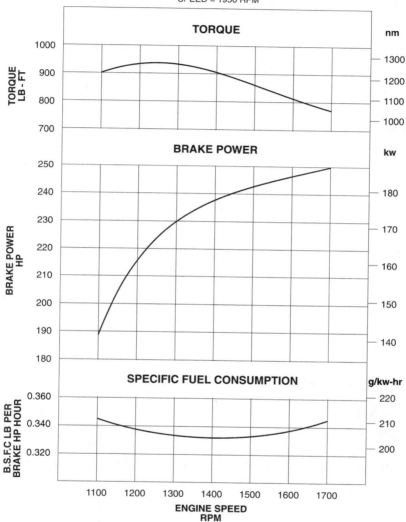

Figure 8-11
Engine Speed
(Courtesy Mack Trucks)

- Torque performance
- Brake horsepower performance
- Specific fuel consumption

You should specify an engine that will operate above maximum torque while under maximum horsepower and on the best fuel consumption area most of the time this vehicle is being used for cost efficiencies.

Further horsepower is needed to overcome air resistance. Using the frontal area measurement in square feet, the following information is an example of rear wheel horsepower needed to push through the air acting as a resistance.

Horsepower Requirements/Frontal Area

Speed		30	35	40	45	50	55	60	65	70	75
Rear Wheel Horsepower Required											
Frontal	50	09	14	21	30	42	56	72	92	114	140
Area	60	11	17	26	36	50	66	86	110	136	169
Sq.	70	13	20	30	42	58	78	100	124	159	197
Ft.	80	14	22	33	49	67	89	115	147	182	225
	90	16	25	39	54	75	99	129	165	204	253

Moving the Load

Moving the load in medium and heavy trucks is not usually much of a problem. Most engine-transmission combinations provide sufficient power and gear reductions for adequate performance. This is not to say you can ignore performance in the selection of components for light trucks, but it is not as critical as in heavier units.

To move the load in a given operation, the vehicle must be equipped with an engine that will furnish adequate power, plus the proper transmission and rear axle ratios to transmit the power to the rear wheels. To obtain maximum performance, several factors must be considered:

- Maximum gear reduction
- Geared road speed

- Gradeability
- Number of gear steps

Maximum gear reduction provides the greatest multiplication of torque in a given power train. This highest reduction is obtained when the transmission is in its lowest gear. Vehicle starting ability is directly affected, and measured, by it. Trucks operated in on- and off-highway service or hilly areas usually require greater reduction than units operating on Interstates highways. As gross vehicle weights increase, greater gear reduction or more power is required.

Geared road speed is the maximum speed with the transmission in top gear and the engine operating at its top governed revolutions per minute. It is limited by gear ratios and governed engine speed. Many vehicles, especially heavier units, cannot obtain this speed because their top speed requires more horse-power than is available.

Gradeability is a measure of the ability of a vehicle to start on and climb a grade. It is indicated by percent of grade. As an example, a 1% grade is equivalent to a 1-ft rise over a distance of 100 ft. A vehicle should have a minimum of 10% gradeability to set the vehicle in motion from a standing start on flat ground. The unit must also have sufficient gradeability to negotiate the steepest grade encountered. For example, if a unit will encounter an 8% grade (a rise of 8 ft in 100 ft of travel) on its normal route, a minimum of 18% (10% starting and 8% operating) gradeability is required.

Speedability is the actual road speed a vehicle can attain on flat ground in still air. Speedability is influenced by the ability of the engine to overcome the combined retarding effects of gross weight, air, and road resistance. Speed-ability should equal or exceed the geared road speed of the vehicle. With today's high gross weight and road speeds, vehicles require significant horsepower (see Figure 8-12) to stay with traffic and get good fuel economy.

Gear steps involve transmission selection. Transmissions vary from 3 to 20 (and more) speeds. Deciding on the right number of gears is important. You could have 20 speeds, but if 7 can do the job, the 7-speed transmission would be preferable because it is easier to drive and more economical.

Transmission and axle ratio selection are related to this and are extremely important. Much attention is required to obtain the right combination. You

Horsepower							
Unit Type, Pounds GVW	At 35 mph	At 40 mph	At 45 mph	At 50 mph	At 55 mph	At 60 mph	At 65 mph
50,000	87	106	128	152	181	214	248
55,000	94	114	137	163	192	226	260
60,000	100	122	145	172	202	236	272
65,000	107	130	154	182	213	247	284
70,000	114	138	163	191	224	259	297
75,000	120	144	171	200	233	271	309

Figure 8-12
Horsepower Requirements at Different Speeds

want the shifts to remain between the top engine torque and the maximum horsepower so that the vehicle will operate according to its engine design but keep up with traffic when loaded. Consulting a chart of engine speed versus geared speed will verify the choice of components. Calculations should be graphed and plotted on the chart.

Example: Figure 8-13 is a sample chart showing the shift patterns calculated for a 7-speed transmission. This particular chart is based on a vehicle equipped with a diesel engine rated at 250 hp at 2100 rpm with 685 ft-lb of torque at 15 rpm, a 7-speed transmission, and a rear axle with a ratio of 5.86:1, and 12:00 × 24 tires. The vertical axis of the graph illustrates engine rpm. The horizontal axis is mph. A horizontal line (in this case, at 1500 rpm) is drawn to indicate the level of maximum torque. Another horizontal line (in this case, at 2100 rpm) should be drawn to show the level of maximum horsepower. Each angled line from 0 to the mph line at 2100 rpm represents the specified gear ratio in which the transmission is engaged. To plot these gear ratio lines, you must first calculate the speed the vehicle will reach each time before shifting to the next higher gear. The following is the formula for calculating each of these speeds:

$$\text{mph} = \frac{\text{governed rpm} \times 0.006 \times \text{rolling radius}}{\text{axle ratio} \times \text{transmission gear ratio}}$$

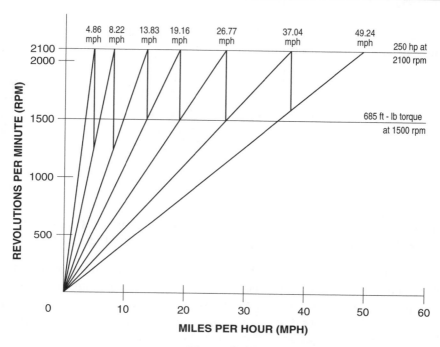

Figure 8-13
Shift Pattern Chart

In this example,

Governed rpm	=	2100 rpm
0.006	=	Conversion factor for friction
Rolling radius	=	22.9 in.
Axle ratio	=	5.86

The transmission gear ratios for the seven gears are as follows:

First	10.13
Second	5.99
Third	3.56
Fourth	2.57
Fifth	1.84
Sixth	1.33
Seventh	1.00

217

Using the preceding formula and figures, you would calculate the top speeds for each gear to be as follows:

First speed	4.86 mph
Second speed	8.22 mph
Third speed	13.83 mph
Fourth speed	19.16 mph
Fifth speed	26.77 mph
Sixth speed	37.04 mph
Seventh speed	49.24 mph

The seven speeds should be plotted along the maximum-rpm line, and these points should be connected to the origin (0 mph) to form the gear lines.

If the proper speeds have been calculated from the proper number of gears for the shift pattern of the vehicle, lines dropped perpendicular from the calculated top-speed points along the maximum-rpm line will intersect the next highest gear line at a point above the torque line, indicating that at this speed there is sufficient torque for this next gear. For example, a perpendicular line dropped from 37.04 mph, the top speed in sixth gear, intersects the seventh-gear line well above the torque cutoff line.

However, notice that similar perpendicular lines drawn from the first- and second-gear lines intersect the next gear at points below the torque line already established. Even the third gear line is marginal. The solution here is to increase the horsepower or the number of speeds of the transmission to extend the service life of this vehicle.

Selecting the Proper Transmission

The function of the transmission in a truck is to translate the horsepower of the engine into the proper speed and torque required to move a given load. For some applications, this can be accomplished with only a 5-speed transmission. Another situation, however, may require 14 or 20 gears to provide maximum low gear for gradeability, close splits for satisfactory performance, and perhaps an overdrive for top road speed. Use the following guide as an aid to select your transmissions.

Five-Speed Gathered Ratio (close fourth and fifth gear). This style transmission is used primarily with single-speed axles for straight city truck service or low GVW highway service.

Five-Speed Equal Step. This transmission is best suited when splitshift is required with an auxiliary transmission or a 2-speed rear axle (Figures 8-14 and 8-16).

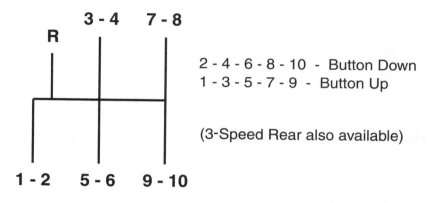

Figure 8-14
Gear Shift Pattern:
2-Speed Rear Axle—5-Speed Transmission

Six-Speed Transmission. This transmission is well suited to work with high torque rise and fuel-efficient engines. Both equal step and gathered ratios are available with direct on sixth gear to use with these engines.

Seven-Speed Transmission. This is the best choice to use for highway service for leasing common carrier or private fleets. Maintenance is reduced because no air is required to make range shifts. It offers a simple shift pattern and good starting gear for the fuel-efficient engines.

Ten-Speed Transmission. This transmission is designed to work in short wheelbase vehicles primarily used in over-the-road fleet operations. The overall length of this transmission (282 in.) makes it ideal for this kind of installation. This transmission will work well with fuel-efficient and conventional engines (Figure 8-15).

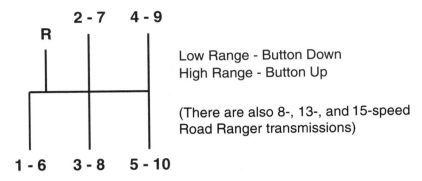

Figure 8-15
Gear Shift Pattern:
Raod Ranger—10-Speed

Thirteen- or Fourteen-Speed Transmission. This transmission is designed for highway and on/off highway applications where versatile gear selection is required to fit all working conditions. This transmission has single stick controls with a four-position splitter valve that can be shifted with a variety of shift patterns, depending on the terrain and type of service.

Twenty-Speed Transmission. The twenty-speed transmission is an excellent selection for 10 speeds off highway or 10 speeds on highway with a progressive shift pattern from Low Hole to highway ratios in first gear position.

Two-Speed, Three-Speed, or Four-Speed Auxiliary Transmission. This unit provides a means of supplementing the gearing in five- or six-speed transmissions. Adequate low gear, close splits, and overdrive ratios are available through the use of auxiliaries (Figure 8-16).

Overdrive Main Transmission. The gathered ratio style transmission is available in an overdrive. The use of an overdrive is often dictated by certain applications or requirements such as being loaded one way and empty on the return trip. This practice can help improve fuel economy by allowing a cruise speed of 55–60 mph at the best position of fuel curve.

In long-haul, high-speed applications, where there would be minimal shifting in these lower gears, this underpowered shift pattern might be acceptable.

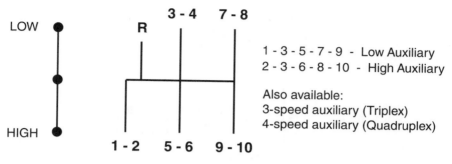

Figure 8-16
Gear Shift Pattern:
Twin Transmission—5-Speed Main—2-Speed Auxiliary

However, if pulling a full load in an urban situation, where a great deal of shifting in the lower gears would be necessary, the engine would tend to be lugged, and the shift pattern would be unacceptable.

There are several ways in which this situation could be remedied. Although raising the maximum rpm would be a possibility, a more likely solution would be either to add more gears between the ones that will cause the engine to lug or to lower the level of torque set as the cutoff. Looked at graphically, any of these alternatives would bring the points of intersection above the torque line.

Gear-Fast Run-Slow Transmissions. The engine speed for cruising in a vehicle is lower than its governed speed. This allows the engine to run at its most economical engine rpm range and the vehicle to run at cruising speed, which is more efficient than governed speed. Governed speed would be higher than an economical engine speed for the same cruising speed.

Clutch Torque Limit Calculations

The lugging forces that would cause premature clutch failures can be calculated.

Clutch Torque Limit (when clutch slips when fully engaged)
7 in. Inside Dia. $T = 0.306\ P\ (r+R)$
11 in. Outside Dia. $T = 0.306 \times (9 \times 130) \times (+11)$
9 in. Spring Pressure Plate $T = 0.306 \times 1170 \times 1.5$
130 lb force $T = 537$ ft-lb Maximum

221

$$T = \text{Torque limit}$$
$$P = \text{Total pressure plate force (ft-lb)}$$
$$r = \text{Inner radius of disc in feet } \} \text{ clutch}$$
$$R = \text{Outer radius of disc in feet } \} \text{ disc}$$

$$TMF = \frac{\text{front counter gear} \times \text{last gear}}{\text{clutch shaft gear} \times \text{last counter gear in use}}$$

$$Torque = \frac{BHP \times RPM}{5252}$$

$$Ratio = \frac{DIA \text{ driver}}{DIA \text{ drive}} = \frac{\text{\# of teeth driven}}{\text{\# of teeth drive}}$$

It is desirable on your part to calculate potential problems to reduce clutch maintenance.

In addition to ensuring that a vehicle has a shift pattern adequate to enable it to maintain road speed, you would also need to calculate gradeability, its ability to negotiate a suitable incline fully loaded at its gross combination weight.

Gradeability is given by

$$G = \frac{K \times M \times R \times T}{GW} - 1$$

where

$$K = \text{Rear axle factor of 0.1001 for a single axle, or 0.095 for a dual axle}$$
$$M = \text{Tire revolutions per mile}$$
$$R = \text{Maximum gear reduction}$$
$$T = \text{Engine maximum net torque}$$
$$GW = \text{Gross vehicle and load weight}$$

Example: Again looking at the truck in the previous examples,

K = 0.095 for the dual axle vehicles
M = 433 tire revolutions per mile for 12:00 ∞ 24 tire
R = 5.86 rear axle ratio × first-speed gear ratio of 10.13
= 59.36
T = 685 ft-lb GW = 50,000 lb

Therefore,

$$G = \frac{0.095 \times 433 \times 49.36 \times 685}{50,000} - 1$$

$$= 33.45 - 1 = 32.45$$

Thus, this vehicle can negotiate a flat surface (requiring a minimum gradability of 10) and negotiate a 22-ft elevation over 100 ft (32 – 10 = 22).

If this vehicle were a dump truck or a cement truck, this kind of gradeability would be desired. Gradeability should be predicated on the type of vehicle environment that is available and the needs of the user. Gradeability diminishes as upshifting takes place. As needed, downshifting compensates for variable environments with the best gradeability being the lowest gear.

Chapter 9

Preparing a Functional Specification for Solicitation

⊹

Preparing a Specification for Solicitation

With the vehicle specification written, your next step is to solicit price bids on your vehicle. You want to create a competitive environment to stimulate the lowest price for your functional specification. Whether you are committed to one make of vehicle or open to various makes, you should communicate your bid to as many bidders as possible within your service territory.

Pre-Bid Conference

To pursue this end, you would ask all vendors to attend a pre-bid conference. This allows you to address any questions in the full presence of all the bidders. You would circulate the functional specification prior to the meeting and review line item by line item the technical items required. The more complex the vehicle, the more time is needed to clarify items and coordinate the chassis, body, and mounted equipment installation process.

With the vehicle chassis, body, and mounted equipment specification presented, all the manufacturers and/or suppliers must be present. A matter of concern is the chassis construction time because the body manufacturer must order, assemble, and mount the body. The final installation of the mounted equipment should be scheduled after or during the body specification review. Body and chassis subframes and the assembly process must be reviewed and coordinated to ensure the proper fit and functionality.

If the mounted equipment is to lift 2000 lb at a horizontal 180°, the chassis and the body frame, the mounted equipment subframe, and the mounted equipment mounting process must be reviewed technically. Will the stresses be acceptable by the chassis frame, the body frame, and the mounted equipment subframe to support the lift weight?

With dump bodies, cement mixers, and plows, the proper chassis strength and construction are required for long life. With a group meeting, the customer will receive input from all present at that time (chassis and body manufacturers shortly thereafter) to ensure that the functional specifications are valid.

Timetable of Deliveries

We must define the format of the bidding process and the timetable of the bidding process. An example of a written format would be:

> "We request that you send the quote to us in one envelope, addressing line item by line item each component that you will supply by brand name or equivalent and a second envelope with the prices defined. Our process will be to evaluate the bid technically for qualification separately from the price to fairly evaluate your offer. If the bid meets the requirements, we will combine that technical evaluation with the quality and business evaluation of your company and the price you quote, awarding the order to the best overall qualified complete package submitted by your organization."

We should define our expectations of delivery of the order and our inspection of the work as it progresses. An example of a written definition of our expectations would be:

> "We will inspect the first chassis when it comes off the assembly line for its conformance to specification prior to being delivered to the body and/or mounted equipment manufacturer no later than 90 days from award of the order. We will pay 70% of the chassis manufacture cost 30 days prior to completion of the entire assembled unit or 60 days after we approve the chassis. At that time, we expect the manufacturer's statement of origin so we can register the entire unit to be put on the road when delivered to our property."

Vehicle Inspection Process

"We expect to inspect the body and/or mounted equipment assembled and operating but not painted. Upon approval of the prepaint inspections, the first unit will be corrected and painted for our final inspection. We will pay for our out-of-pocket expenses to accomplish this. Should the first unit not be satisfactory at the final painted inspection, we would expect the manufacturer of the body and/or mounted equipment to pay for our out-of-pocket expenses at that time and for our future visits. Our time is to be at $50 per hour, or whatever is appropriate from departure of the home office to return, plus out-of-pocket or an agreed upon payment."

This point must be clarified so that the customer is not used as an inspector. You should intend and want the manufacturer to use your specification as a guide to inspect its work to your standards.

Notification of Inspection Dates

You must define the notification process of the prepaint inspection date three weeks prior, and the delivery date must be defined so that the customer can prepare for the vehicle acceptance process. When the vehicle is delivered, you should inspect it, perform an acceptance preventive maintenance procedure, and decal and install items for service (fire extinquishers, transfer of stock, etc). After the vehicle is prepared for in-service, you should quantify an acceptance time for in-service check to ensure that all accessories work properly.

For example:

"The unit will be put in service two weeks after delivered on site to the customer, and the shakedown period will be 30 working days after in-service. After the 10-day in-service and 30-day shakedown period, the unit, if satisfactory, will be approved for payment. A check will be generated 20 working days after acceptance, which is sixty (60) days after delivery to the customer's property..."

"Liquidated Damages" Correcting Vehicle Problems

Should the unit be troublesome and the problems not corrected during the 30-working-day shakedown period, it is desirable to quantify action taken.

For example"

> "Should the vehicle not be approved for service 40 days after delivery, payment will not be authorized. If the unit is not corrected 30 days after the 30-day shakedown period, which is 70 days after delivery, a $3,000 per month penalty (or whatever penalty dollar is appropriate to offset the increased expense the old unit is costing to keep it in service) will be subtracted from the invoiced amount (liquidated damages)..."

Warranties

Warranty expectations should be defined. Standard warranties are supplied by the manufacturer, or you can quote your own to support your needs, (e.g., online communication versus paper transactions).

For example:

> "All work will be warrantable for full labor and parts for one year after final acceptance and payment by the customer. Each repair completed during this period will be warranted for two full years from date of completion and acceptance by the customer."

Article Warranty

1. The Contractor acknowledges that he is an expert fully competent in all phases of the services required under this contract. The Contractor warrants that he shall, in good workmanlike manner, perform all work and furnish all supplies and machinery, equipment, facilities, and means, except as herein otherwise expressly specified, necessary or proper to facilities and means, except as herein otherwise expressly specified, necessary or proper to perform and complete all the work required by the contract. Although the Owner may review and approve various documents including, but not limited to, Quality Assurance Manuals, Installation Agreements, and so

forth, the Owner neither accepts in any manner responsibility for nor relieves the Contractor from responsibility for the performance of all the requirements of this contract. The Contractor warrants that all material furnished will be free of any defects in material and workmanship, and will be the kind and quality as specified by the purchaser. If any failure to meet the foregoing warranty appears within one (1) year after the date first placed in use, the Contractor will correct by repair or replacement and provide that each such repair or replacement will carry this same warranty for two (2) full years starting upon completion of said repairs or replacement. The Contractor shall be responsible for all removal costs into and out of the location(s) the Owner has designated wherever a repair or replacement is required.

2. The foregoing shall not negate any other warranties of the Contractor, expressed or implied by operation of law.

Article—Latent Defects

Notwithstanding any other clause in this contract, should a "Latent Defect" in the work be discovered during the five (5) year period after the work has been accepted by the Owner, it shall be the Contractor's responsibility to repair or replace the work. A "Latent Defect" is defined as a design defect that reasonably careful inspection will not reveal, or a defect that could not have been discovered by inspectors, or a defect that could not be located by any known or customary test.

Liquidated Damages

If the unit is not fully accepted when delivered for preliminary acceptance at the Company Fleet Garage and such delay is the responsibility of the dealer or body manufacturer, all costs for a compatible vehicle rental will be paid by the chassis or body vendor responsible. If the vehicle is not turnkey at the end of 60 days from delivery acceptance at the Company and fully return all monies within 30 days to the Company Fleet Garage, then the vehicle will be returned at no cost to the Company.

Additional Issues

When the vehicle is delivered, it will be weighted by axle from a certified scale empty, compared to CAD-CAM design information for acceptance.

Latent Defects

Another area of concern is vehicle latent defects. These are design failures and failures due to poor workmanship by the manufacturer or the subcontractor supplying the manufacturer. An example would be a frame failure due to excessive rust or poor quality metal or alloy blend.

Usually each state quantifies latent defect periods—customarily, it is four years in length. Your expectations must be quantified in this area, usually exceeding the state requirements to cover failures. This keeps the door open for your valid claims process should it be needed, because some expectations exceed those of the state, and the vendor agrees to this item.

Costs to Vehicle Bids

These previous items could cost you extra in the price of the vehicle. The vendors must understand the terms of your bid requirements so they can bid accordingly. With the competitors established, each vendor will carefully tender its proposal and moderate extra cost inclusions.

If transportation has had success with vendor delivery, and small, medium, and/or heavy vehicles with no in-service problems, transportation should be cautious about unnecessary requirements. In most cases, due to people process and material problems by the manufacturers (similar to people process and material problems), it is desirable to quote and define your expectations to ensure a timely and cost-effective in-service process. The more detail in your specifications, the more room for a misunderstanding to take place. Thus, a pre-bid request for proposal meeting should take place to ensure a full understanding of the customer requirements.

This is fair for both the manufacturer and the customer.

Standard Attachments

The following is an example of some "boiler plate" or standard terms and conditions language attached to your specification to ensure that it will cover your needs and expectations clearly and completely, so the bidders can tailor their bids to your needs at the lowest, most competitive price.

Date: _____

Bid #: _____

Subject - RFP No.: _____

For: _____

Gentlemen:

You are invited to submit your proposal to Company for the above-referenced services. Your Proposal should be in full accord with these RFP requirements, including the following documents:

1. Proposed Contract #_____
2. Scope of Work # _____
3. Instruction to Bidders and Proposal Forms _____
4. Bidders Response Notice _____
5. Specification # _____
6. Drawing #_____
7. Green envelope/label for sealed proposal; Blue envelope/label for specification line item, proposal.
8. "General Provisions" - Revision F dated 1 October 2002 including Appendix "A," Insurance Requirements
9. Reference Documents

The Proposal due date is not later than 4 p.m. on _____

To be delivered care of RFP# _____ at our offices in _____.

Bidders are requested to notify J. Jones regarding their intention to respond to this inquiry by completing the attached notice. (212) xxx-xxxx

Your interest in this effort is appreciated, and you may be assured that your proposal will receive prompt and careful attention.

Send to: J. Jones
             Company
             Address

INSTRUCTIONS TO BIDDERS AND PROPOSAL FORM

1. GENERAL

   The Draft Contract Form and attachments thereto describe the work to be performed and represent the form and terms of X Company's anticipated agreement with the contractor selected for this project.

2. PROPOSAL SUBMISSION

   One complete proposal must be forwarded, utilizing the enclosed green sealed bid envelope/label, to

   > J. Jones
   > Company
   > Address

   One separate copy of the proposal form and schedule information, less final price information, Blue envelope/label should be forwarded directly to:

   > J. Jones
   > Company
   > Address

   Your proposal must be received by the time and date specified in the inquiry transmittal and must be firm and not subject to change for a period of sixty (60) calendar days after the proposal due date.

3. PROPOSAL RESTRICTIONS

   Any proposal received after the aforementioned time and date may be retained by J. Jones, but J. Jones has no obligation to either evaluate or return such delinquent proposal. X Company reserves the right to postpone the time and date for submission of proposals at any time prior to the proposal deadline by giving written notice of such postponement to each prospective bidder.

   Failure by the bidder to fill in all blanks of the proposal form and to supply completely all information required may result in the proposal being rejected by X Company at its option. Any exceptions to the inquiry documents may be a basis for considering a proposal to be non-conforming.

INSTRUCTIONS TO BIDDERS AND PROPOSAL FORM (CONTINUED)

Submission of a proposal shall constitute acknowledgement by the bidder that it has thoroughly examined all documents that are part of the inquiry, including all addenda that may be issued during the proposal preparation period. No claim will be allowed for additional compensation or additional time for completion which is based on lack of knowledge or lack of understanding of any document. Additionally, any communication from a bidder which in any manner discloses price information contained in its proposal, if received prior to opening, may be cause for disqualification.

No bidder may withdraw his proposal after the hour set forth for opening proposals. All proposals are to remain in effect for sixty (60) days. Any and all costs incurred by the bidder in conjunction with furnishing a response to this inquiry shall be at the bidder's expense and borne by the bidder.

4. BASIS OF AWARD

    X Company reserves the right to make an award of the contract based on responses to this inquiry or to negotiate with any or all of the bidders to the best advantage of X Company. In making an award of the contract, X Company reserves the right to consider factors other than price such as past experience, management capability, schedule compliance, etc.

5. REJECTION OF BIDS

    X Company reserves the right to reject any or all bids.

6. RETURN OF INQUIRY DOCUMENTS

    Those bidders who decline to quote or who are not awarded a contract are requested to return all inquiry documents to X Company c/o J. Jones.

PROPOSAL FORM

1. BIDDER'S STATEMENT

    _____ (firm) submits the following proposal for performing all work required by, and as specified in, the Inquiry Documents for Company's Inquiry Number _____ .

2. PROPOSED PRICE

    Complete attached "Price Proposal Form" and forward with one copy to: ATTN: J. Jones, Contract Department, using the green sealed bid label. Forward one copy of your proposal <u>without pricing data</u> using the blue sealed bid label to ATTN: Mr. J. Jones, Contracts Department.

3.. ADDITIONAL INFORMATION

    A. List all proposed subcontractors, nature of work, and percentage of work in terms of contract amount to be performed by each subcontractor:

        Subcontractor/Address   Type of Work   Net %

    B. Provide the following financial data current as of the latest completed fiscal year:

        Fiscal Year Ending: _____   Audited By: _____
        Net Sales: _____   Net Profit: _____
        Net Assets: _____   Net Liabilities: _____
        Current Assets: _____   Current Liabilities: _____
        Net Worth (Equity): _____

    C. PERFORMANCE SCHEDULE

        Complete the attached "Performance Schedule" form indicating the number of calendar days after contract award required for completion of events that are significant to the performance of this contract.

    D. PAYMENT SCHEDULE

        If the payment terms of this contract provide for payment by X Company on a "Milestone Payment" basis, attach a list of proposed milestone events and corresponding payment amounts.

    E. INSURANCE

        Attach insurance certificate in accordance with insurance requirements (See General Provisions).

    F. EXCEPTIONS

        Bidder takes the following exceptions to X Company contract requirements as expressed in the inquiry documents (if none, so state): _____

    G. AUTHENTICATION

        This proposal shall be valid for a period of sixty (60) calendar days from the date established by _____ for submission of proposals.

By: _____   _____
    (Authorized Signature)   (Name of Firm)

Name: _____   _____
                                     (Address)

Telephone: _____
                                         (Affix Corporate Seal)

PERFORMANCE SCHEDULE

Inquiry No.: _____

Submitted by (Name of Firm): _____

The Work Schedule shall commence upon Contract Award. Provide the days to complete for:

Significant Events	Completion By: (Calendar Days After Award)
Contract Award	X*
1.	X + _____days
2.	X + _____days
3.	X + _____days
4.	X + _____days
5.	X + _____days
6.	X + _____days
7.	X + _____days

*All events must be scheduled from the starting date of Contract Award which is designated "X."

NOTE: Any restrictions regarding the award date by X Company must be stipulated in your proposal. Failure to so stipulate will indicate that upon award by X Company that this performance schedule can be met by the Contractor.

PRICE PROPOSAL FORM

Inquiry No: _____

PROPOSED PRICE

Lump Sum Price for furnishing all materials, labor, equipment, insurance, and all other items specified in and required by the Inquiry Documents (except those items specifically excluded), and fully performing in a good workmanlike manner the work set forth therein:

_____dollars	($_____)	which consists of:
_____dollars	($_____)	for material and
_____dollars	($_____)	for labor

This price proposal also consists of the following Unit Price Sheets and Time and Material Rate Sheets which pertain to work added to or deleted from the Contract Scope of Work.

This proposal shall be valid for a period of sixty (60) calendar days from the date established by for submission of proposals.

By: _____   _____
      (Authorized Signature)   (Name of Firm)

Name: _____   _____
                                                                      (Address)

Telephone: _____   _____

                                                                   (Affix Corporate Seal)

235

Contract No. _____
Date _____

FIXED PRICE CONTRACT

THIS AGREEMENT MADE AS OF THE _____DAY OF _____2002

Between _____
_____
_____
subsidiary company of the hereinafter referred to as "Owner"

And _____
_____
_____
hereinafter referred to as "Seller"

WITNESSETH THAT:

IN CONSIDERATION OF the premises and mutual covenants and agreements herein contained, the parties hereto agree as follows:

1. ENTIRE AGREEMENT
   This contract supersedes all written or oral agreements, if any, and constitutes the entire agreement between the parties.

2. SCOPE OF WORK
   The Seller shall furnish all the services and materials necessary to complete the work set forth in Attachment 'B.'

3. DELIVERY
   Delivery Schedule - The items, services, and deliverable data required by this contract shall be delivered per the following:
   a. Time is of the essence relative to delivery of units under this contract.
   b. Seller shall contact _____ Transportation Department S. Smith (Mgr. Transportation, (212)XXX-XXXX) for delivery location thirty (30) days in advance of delivery. Unit shall be drop ship by manufacturer.
   c. Delivery shall be F.O.B. Destination, Freight Allowed.
   d. Delivery must be made on or before _____ or ninety (90) days after receipt of chassis.
   e. Vehicle Delivery Inspection Program

   Scope
   This policy applies to all company vehicles including sedans, light trucks, heavy trucks, and work equipment delivered.

   Policy
   Corporate Transportation is responsible for coordinating a vehicle inspection of all new vehicles or vehicular work equipment used in the company's operations.

FIXED PRICE CONTRACT (CONTINUED)

Sedans and Light Trucks

Inspections of sedans and light trucks that are mass produced and delivered to the company facility or through a manufacturer's dealer organization are to be performed within a 24-hour period after delivery.

Work Equipment and Mounted Equipment

Inspections of work equipment and trucks with bodies or mounted equipment are to be performed at two or more separate times in their production and delivery.

The first inspection should be performed just prior to painting of the equipment. At this time, vendor compliance with the written specifications is to be made. The vendor's representative is to prepare a list of items identified during the meeting. A copy of this list is to be taken by the inspector for review at the final paint inspection. Any minor no-cost changes or resolution of details may be resolved prior to this final inspection.

Changes that will affect the vendor's quoted price and other disputed items that cannot be resolved at the prepaint meeting will be brought to the attention of the transportation Buyer and the Manager-Transportation for resolution with the vendor.

Additional inspection at vendor's cost due to excessive items left uncorrected at final inspection may be made prior to shipment from the vendor's plant to insure that all specifications are met. Such inspections may include, but are not limited to, a review of vehicle and equipment stability under test loads, hydraulic system operating characteristics, paint quality and finish, wiring, assembly, and electrical system performance.

An acceptance inspection of mounted or work equipment is to be performed at the time of delivery to the company property or within a 10-working-day period after delivery.

Should there be any discrepancies during this 10-working-day period, they must be completed before an in-service period starts. A 30-working-day in-service trial commences at the end of this 10-working-day period. During this 30-working-day in-service trial period, any discrepancies uncovered must be corrected. At the end of the 30-working-day trial period when the unit is accepted, payment in full will be initiated within 20 working days.

4. AUTHORIZATION FOR CHANGES AND APPROVALS

No changes or amendments to this Agreement are authorized unless made by the X Company representatives designated by name below and substantiated by formal written direction or amendment.
   1.   J. Jones - Contract Manager
   2.   P. Tello - Manager of Contracts

IN WITNESS WHEREOF, the parties hereto have executed this contract as of:

Date: _____   Date: _____
For: _____   For: _____
By: _____   By: _____
Title: _____   Title: _____

SCOPE OF WORK

1.0 Article to be Delivered

    1.1 Seller shall provide all facilities, skills, services, and materials necessary for and incidental to the design, development, fabrication, test, and delivery in accordance with the requirements set forth in this contract and all exhibits, specifications, and drawings incorporated by reference and made a part hereof.

    1.2

	Quantity	Description
1.2.1	_____ Each	_____
		_____
		Technical Specification
		No._____
1.2.2	_____ Sets	Manufacturer's Service Manual and service specification (two for each vehicle).

2.0 Services to be Performed

    2.1 Provide vehicles to meet specification requirements for final acceptance by the Company.

    2.2 Seller shall maintain a Quality Control Program as required in Section 5.2 of ANSI.

        2.2.1 QA records shall be maintained on an individual vehicle basis.

        2.2.2 QA records shall be made available to personnel at time of acceptance of vehicle.

    2.3 Upon receipt of chassis, body manufacturer will perform inspection to insure unit(s) are not damaged and are correct for intended use. Forward receiving inspection report to Mgr. Transportation (212) xxx-xxxx.

3.0 Payments

    3.1 Payment will be made through a leasing arrangement with Leasing Corporation. Partial payment (70%) of chassis will be agreed to (30 working days before delivery to X Company) for title/MSO for license plates.

    3.2 The Lessor authorizes and appoints X Company to enforce, in its own name, any claim, warranty, agreement, or representation that may be made against any supplier of cab and chassis, bodies, or other equipment associated with this order.

    3.3 Registration and plates to be obtained by X Company 30 days prior to delivery.

    3.4 Original invoice shall be sent to J. Jones

    3.5 Final payment will be authorized by J. Jones, X Company, 40 working days after delivery of unit to X Company Transportation Department.

4.0 Acceptance

    4.1 Any item purchased hereunder shall be subject to inspection and test by Owner to the extent practicable at all times and places including the period of manufacture and, in any event, prior to final acceptance. Owner may inspect the plant or plants of Seller or of any of its subcontractors engaged in the performance of this contract.

    4.2 If any inspection or test is made by Owner on the premises of Seller, Seller, without additional charge, shall provide all reasonable facilities and assistance for the safety and convenience of Owner's inspectors in the performance of their duties. All inspections and tests shall be performed in such manner as not to unduly delay the

SCOPE OF WORK (CONTINUED)

work. No inspection or test made prior to final inspection and acceptance shall relieve Seller from responsibility for defects or failure to meet the requirements of this contract.

4.3   Final inspection and acceptance of items shall be made by Owner after delivery and shall be conclusive except as regards latent defects, fraud, such gross mistakes as amount to fraud, and Seller's warranty obligations.

4.4   Seller shall provide and maintain an inspection system in accordance with sound business practices and as otherwise provided in this contract. Records of all inspection work by Seller shall be kept complete and available to Owner during the performance of this contract and for three (3) years after final payment.

4.5   Said representative shall have full access to vehicles, all tests, inspection records, and reports associated with vehicles required to determine that all requirements of the specifications have been satisfied.

4.6   Said representative shall have the right to exercise or have exercised all moving parts of the vehicles to demonstrate to his satisfaction the acceptable function of all parts.

4.7   Representative shall authorize release for shipment, and final acceptance will be made at delivery destination.

4.8   Final acceptance/rejection will be accomplished in the following manner:

   4.8.1   Receipt of vehicles will be acknowledged by an X Company representative — initiating a 40-day in-service debugging period.

   4.8.2   Operating test shall be performed on vehicles under in-service conditions. Corrections and/or adjustments required by vehicles to obtain performance in accordance with specifications shall be recorded.

   4.8.3   If malfunctions occur which cause vehicles to be placed in an out-of-service condition, Seller shall be notified immediately and repair or replacement shall be made by Seller to return vehicle to in-service condition. The downtime caused by the out-of-service condition shall be added to the in-service debugging period on a day-for-day basis to provide a full 40-working-day debugging period.

   4.8.4   At completion of the 40-working-day in-service debugging period, a final inspection and test shall be made of the vehicles. The test will be completed in two days.

   4.8.5   If at completion of final inspection and test vehicle is considered acceptable, Seller shall be so notified, and Lessor shall be informed of acceptance and authorized to make payment within 20 working days.

   4.8.6   If at completion of final inspection and test vehicle is not accepted, a punch list defining rejection, corrections, and/or adjustments to be made will be presented to Seller for corrective action.

   4.8.7   Seller at his cost shall return vehicle to point of manufacture, complete corrective action, and redeliver vehicle to X Company. If agreeable with X Company, Seller may complete corrective action as a field repair.

SCOPE OF WORK (CONTINUED)

4.8.8 Upon correction of all items on punch list, X Company shall reinspect and test punch list items to complete final acceptance within two days after resubmittal of vehicle for inspection and acceptance.

4.8.9 Title shall pass upon preliminary acceptance and inspection by X Company's representative.

4.8.10 If unit is not fully accepted within 60 working days of preliminary acceptance at X Company and such delay is the responsibility of dealer or body manufacturer, all costs incurred by X Company for late payment or interest payment on the vehicles will be charged to the account of the dealer or body manufacturer causing the delay. Such charges shall include average cost of leasing and maintenance (established at $3,000 per month) of proposed vehicles to be replaced or retired.

5.0 Warranties

5.1 Contractor warrants that all material furnished will be free of any defects in material and workmanship, and will be the kind and quality as specified by the purchaser. If any failure to meet the foregoing warranty appears within one (1) year after the date first placed in use, Contractor will correct by repair or replacement and provide that each such repair or replacement will carry this same warranty for two (2) full years starting upon completion of said repairs or replacement. The Contractor shall be responsible for all removal costs into and out of the location(s) Owner has designated wherever a repair or replacement is required.

5.2 The foregoing shall not negate any other warranties of Contractor, expressed or implied by operation of law.

6.0 Labor Disputes

6.1 Whenever an actual or potential labor dispute is delaying or threatens to delay the performance of the work, the Contractor shall immediately notify Owner in writing. Such notice shall include all relevant information concerning the dispute and its background and the steps proposed by Contractor to resolve the dispute or prevent its occurrence.

6.2 In the event of a work stoppage or labor dispute, involving the Contractor or his subcontractor(s), the involved contractor shall, on the request of the Owner, promptly initiate proceedings in such administrative, judicial, or arbitral forum having jurisdiction to resolve, or minimize the impact of, the work stoppage or labor dispute.

7.0 Latent Defects

7.1 Notwithstanding any other clause in this contract, shall a "Latent Defect" in the work be discovered during the five (5) year period after the work has been accepted by the Owner, it shall be the Contractor's responsibility to repair or replace the work. A "Latent Defect" is defined as a defect that reasonably careful inspection will not reveal; or a defect that could not have been discovered by inspectors; or a defect that could not be located by any known or customary test.

SCOPE OF WORK (CONTINUED)

8.0 Insurance Requirements

A. Contractor represents that it now carries, and agrees it will maintain, Workers' Compensation, Employer's Liability, Comprehensive General Liability, and Auto Liability Insurance at the following minimum limits:

			Limits
1.	a.	Workers' Compensation Insurance	Statutory
	b.	Employer's Liability Insurance	$500,000
2.		Comprehensive General Liability (Public Liability) Insurance including:	
	a.	Bodily Injury and Property Damage	$1,000,000 per occurrence or per claim
		or	or
	b.	Bodily Injury and Property Damage	$500,000 per occurrence or per claim and $1,000,000 combined single limit per occurrence or per claim
	c.	Blanket Contractual	Included
	d.	Products and Completed Operations Hazard	Included
	e.	Broad Form Property Damage	Included
	f.	Personal Injury	$500,000 per occurrence or per claim

NOTE: If any of the work performed under this Contract includes blasting; excavating; pile driving; caisson work; moving, shoring, underpinning, razing, or demolition of any structure; removal or rebuilding of any structural support thereof; or any subsurface or underground work, the Comprehensive General Liability Insurance policy shall include coverage for the explosion, collapse, and underground hazards.

3.		Automobile Liability Insurance (owned, hired, and non-owned):	
	a.	Bodily Injury	$500,000 Per Person
			$500,000 Per Accident
	b.	Property Damage	$100,000 Per Accident

B. During the course of construction and until work is finally accepted, and at the discretion of Owner, Contractor may be required to provide and maintain Fire, Extended Coverage, Vandalism and Malicious Mischief Insurance, covering Owner and Contractor as their interest may appear on the buildings, structures, machinery equipment, supplies, and structures of all kinds and their contents incident to the construction. When not otherwise insured, Contractor shall carry Fire and Extended Coverage insurance on construction machinery, tools, and equipment belonging to the

SCOPE OF WORK (CONTINUED)

Owner or similar property belonging to others for which the Owner may be liable. All of the above-mentioned property is to be insured while forming a part of or contained in any buildings or structures mentioned above or while in the open on the premises, except that such insurance may exclude property buried in the ground or noncombustible property in the open.

1. Contractor and its subcontractors shall carry and maintain at their own expense Fire and Extended Coverage insurance on their machinery, tools, equipment, and clothing belonging to them or their employees.

2. Contractor shall provide a waiver of any rights of subrogation which the Contractor may have against Owner, its agents, or its employees.

3. Before any of the work is started under contract, the Contractor shall file the Company certificates of insurance containing the following information in respect to all insurance carried:

    (a) Name of insurance company, policy number, and expiration date.

    (b) The coverages required and the limits on each, including the amount of deductibles or self-insured retentions.

    (c) A statement indicating that the Owner shall receive thirty (30) days notice of cancellation or modification of any of the policies that may affect Owner's interest.

    (d) The Owner as an additional insured (except Workers' Compensation insurance).

    (e) If a vehicle is carrying your hazardous waste, the certificate must show that the vehicle is insured for limits specified in Motor Carrier Act of 1980, as amended.

C. Before permitting any subcontractor to perform any work under this Contract, the Contractor shall require such Subcontractor to furnish certificates evidencing that it now carries and agrees to continue to carry insurance in strict accordance with the requirements of paragraph A of this Article. The Contractor shall maintain a file of such certificates which may be examined by the Owner at any time.

D. The above listed coverages are minimum requirements. If the project is of such a size or nature that it is in the Owner's best interests to require and increase or authorize a reduction in the insurance requirements, it will be done at Owner's discretion prior to the awarding of a contract.

Summary of Vehicle Solicitation Processes

The functional specifications and the contract, scope of work, drawings, and blue and green labels are then prepared and sent to the bidders list for bidders to submit their bids.

Generally, this process can be added to or deleted from, based on past experiences with vendors. This is intended to create an environment for a company to evaluate responses with leverage stated upfront for good business practices.

Should any discrepancies occur, the company can operate within an agreed format to negotiate and resolve differences both perceived and real to both parties' satisfaction.

Chapter 10

Vehicle Bid Analysis

Vehicle Bid Analysis

The process of an efficient vehicle purchasing program using vehicle specifications, vehicle life-cycle costing, and vehicle bidding and the bid analysis is a complex, busy, technical process, but it is not necessarily complicated. It is a <u>necessary</u> cost-effective effort in today's competitive fleet management environment.

Ours was an efficient agricultural society until the 1900s. Maintaining that agricultural efficiency and entering an industrial society was initiated by machinery development and Henry Ford's application of people and machinery in the early 1920s through assembly line manufacturing. Today, due to labor costs, South America, Asia, and the East are becoming increasingly competitive in replacing our labor force, and we are becoming an administrative society. Now our expertise has focused on the analyzing of cost-effective alternatives of statistical data gathering that generates accurate past cost information. Upon that base, we can predict future cost information from these historical details.

Data Gathering

Accuracy has been improved through data processing programs, but data gathering and input still have to be managed for good detail, timeliness, and efficient information turnaround time to ensure a good base of information for an accurate analysis.

Initially, vehicles are specified by component as to what is wanted in a vehicle. You note physical items and/or brand name and/or specific line items with which you had previous successes, such as:

> 125-amp Delco alternators
> Stop master air wedge brakes
> 5-speed Clark manual transmission, 255 cu. in. 150-hp engines
> All-weather tread tires
> 14-ply steel belted radial regroovable and recapable casings

The reason for this approach is that historical information has identified lowest cost per component life cycles for these units. A hypothetical example is a 125-amp alternator, although a higher specification might be needed. Past maintenance history shows that 65-amp alternators have been replaced every 30,000 mi, 90-amp alternators every 60,000 mi. Thus, a 250,000-mile 125-amp alternator is a most cost-effective application of an overspecified item, based on past utilization cost.

To verify this hypothesis, you should test the life of each component or learn from others to verify the concept. Applications change in each fleet, and similar situations breed different results. Thus, you should and must test it yourself in your own application before you believe the cost information that is identified.

Is the Lowest Bid the Best Bid?

When it is time for a fleet manager to replace a vehicle, it is time to upgrade the capabilities of the vehicle that is being replaced. When this happens, the internal user, fleet manager, external customer, and company all benefit.

Let us say the internal customer has a vehicle close to ten years old that is used by his crew. Perhaps it is a rebuilt vehicle with a new chassis put under a rebuilt body and mounted equipment. However, work methods have changed, crew sizes have become smaller, and more tools and accessories are available to increase the jobs this truck can do. Hence, a new truck will have an increased use and productivity, possibly allowing you to replace two older trucks with one new vehicle. One example of the two-for-one concept is the replacement of a digger truck and a derrick truck with a digger/derrick truck. Still not convinced? Today, a single articulated material-handling double bucket truck can remove and replace a transformer (rather than have both a bucket truck and a derrick work together).

The internal customer should see new model development, technology changes, and adjust and change the work methods, allowing more choices to start the innovative, out-of-the-box ideas that could improve the operation of the company. The fleet manager sees these changes and upgrades through the supplier's productivity at trade shows, assembly plants, repair shops, work-shops, seminars, and technical magazines such as *Utility & Telephone Fleets*. Innovation and out-of-the-box ideas cost money to develop. Our suppliers take the time, money, and developmental skills to reduce these expenses and develop a turnkey product for their customers' benefit. There is a gestation period, or trial-and-error period, that the supplier passes on to the fleet managers, the company, and the customers of the company for their benefit. This relationship is critical in establishing an efficient change process.

The fleet manager has a maintenance and operating cost/expense audit trail for all equipment and vehicles, broken down by component, year, class, and application (on-off road, city, suburban, rural) for historical review. When change is agreed upon as an initiated action item, the supplier and fleet manager can exchange pertinent data to explore as a means of making a better product; a product that is more reliable, cost-effective, and able to facilitate a work method improvement.

Low Bid Versus Best Bid

- Define terms and conditions
- Cost of living changes
- Material increases/credits
- Labor increases/credits
- Warranty range—None versus standard versus extended
- Extended warranty on warranty repairs
- Quality of components—Repair costs
- Average bid versus low bid "extremes"
- Defects 5% then change out all vehicles/components
- Recalls 10% then change out all vehicles/components
- Resale—Buy back where appropriate

The fleet manager and supplier must have a synergy, a partnership, and a mutual interest to foster improvements. The supplier is an important ingredient. Having a symbiotic relationship with the supplier will benefit all parties.

Manufacturers, distributors, and suppliers must communicate to maximize life-cycle periods and product improvements. However, those players are competitive when it comes to price because when a vehicle is spec'd and put out for bids, they assume that the low bid will be the winner. However, the lowest bid is not always the best bid.

Technical specifications are detailed to the most minute degree—nuts, bolts, system, and component specifications.

Performance specifications are our specifications. We must determine our needs and hold the supplier liable for performance claims.

The technical specification places the liability on the specification writer. With technical specifications, the low bid is the best bid because all bidders must supply an application-specific vehicle.

The performance specification allows the bidders to distinguish their products from the competition. The difference makers are the capabilities or value-added benefits that could generate cost benefits for the user. As fleet managers, we usually should discard low and high bids and evaluate the several bids in the middle for best bid acceptance.

Yes, you can have change without improvement, but you will not have improvement unless you have carefully planned change. The utility fleet manager is the author of carefully planned change.

Interpretation of Information

Figure interpretation can vary, and statistics can be misinterpreted easily due to varying databases, misapplication, neophyte management, convert enthusiasm, and marketing information. These are only some reasons to approach with caution the use and interpretation of statistics.

Once you prepare the vehicle specifications, you should go out on a Request for Quote to verify vendor response. To ensure maximum competition, you might change specifications to allow more competitive bidding such as:

125-amp alternators or equivalent

Wedge-type air brakes or equivalent

5-speed full synchromesh transmission with a 600-ft torque minimum on input shaft or equivalent

All-weather type tread tires with 14-ply steel-belted regroovable and recapable casings or equivalent

Request for Quotes (RFQ)

The RFQ, tailored for a maximum response, discussed and reviewed with vendors prior to solicitation, now goes out as an RFQ so that the vendors are not surprised with a bid request. Vendors now receiving it, having already discussed your objectives with you, will know better your needs and will be inclined to respond to the bid request in detail. You will have encouraged more vendors due to your pre-bid discussion and their better understanding of your needs to bid for your specification, thus increasing competition, potentially lowering price, and ensuring a quality product.

The reason to spend the time communicating on specifications here is to expand your vendor selection for bid evaluation. To discourage qualified vendors limits competition, which theoretically raises prices and negates your evaluation procedures.

The bids now are reviewed for "apples to apples" response. The underspecified component is penalized for not meeting minimum requirements, and the over-specified component is given additional consideration for your needs. The technical and commercial items are tallied and summarized by the Transportation Department, without the knowledge of vendors' prices.

A suggested weighted review is:

> 40% Technical Review—Line item for line item
> 20% Commercial—Past business experience
> 40% Price Review—Lowest total dollar

Price is the simplest to explain. The lower dollar unit is rated higher than the more expensive, based on an apple-to-apple bid response.

This will be factored in with the next two areas. A matrix evaluation will be used because of a functional specification format. If you were evaluating a technical specification, you would consider only price. Here, price, technical, and commercial elements will be considered for a final cost-effective choice.

Technical Evaluation

The first area to be reviewed is the technical area. You identify the technical areas to be considered and assign them a weighted percentage for evaluation. The weight factor is defined on a numerical level:

5 is excellent
4 is good
3 is acceptable
2 is fair
1 is poor
0 is unacceptable and disqualified

Each vendor is multiplied out item for item, and points are totalled. After points are totalled, you divide each total by the largest total, normalizing the evaluations to 1.

TECHNICAL REVIEW FORMAT

Elements	Weight Factor	Vendor A Rate	Score	Vendor B Rate	Score	Vendor C Rate	Score
Conformance to Spec	30	4	120	5	150	4	120
Quality of Product	20	4	80	5	100	4	80
Service Maintenance	5	5	25	5	25	4	20
Product Reliability	20	4	80	4	80	4	80
Spare Parts Support	10	5	50	4	40	4	40
Technical Support	5	4	20	4	20	4	20
Ease of Maintenance	5	5	25	4	20	4	20
Service of Manual Quality	5	4	20	4	20	4	20
	100		420		455		400

Ratings Defined   5 - Excellent   4 - Good   3 - Acceptable   2 - Fair   1 - Poor   0 - Unacceptable

You summarize the vendors by normalizing their totals. This is accomplished by dividing each total by the highest vendor total and equating each to 1.00.

Vendor Summary	Score		Normalized Factor	=	Final Score
Vendor A - Technical	420	-	455	=	0.92
Vendor B - Technical	455	-	455	=	1.00
Vendor C - Technical	400	-	455	=	0.88

Commercial Evaluation

The next area is the commercial area, and this evaluation is normalized to 1.

The commercial area would be a similar evaluation format. This evaluation is initiated if the bid is technically acceptable.

COMMERCIAL REVIEW FORMAT

Elements	Weight Factor	Vendor A Rate	Score	Vendor B Rate	Score	Vendor C Rate	Score
Warranty Compliance	10	5	50	5	50	4	40
Field Service	10	4	40	5	40	4	40
Inventory Effect	5	5	25	5	25	4	20
Training Technical	5	4	20	4	20	4	20
Training Operating	5	5	25	4	20	4	20
Geography	10	4	40	4	40	4	40
X Delivery Performance	15	5	75	4	60	4	60
X Quality Program	20	4	80	4	80	4	80
X Follow-up	-						
X Manufacturing Cap.	10	5	50	4	40	4	40
X Management Support	5	4	20	4	20	4	20
X Conformance to Bid	5	5	25	4	20	4	20
	100		450		415		400

X - Information from Contracts.
 (Contracts fills in this area after Transportation evaluates the first six areas.)
Ratings: 5 - Excellent   4 - Good   3 - Acceptable   2 - Fair   1 - Poor   0 - Unacceptable

Vendor Summary	Score		Normalized Factor	=	Final Score
Vendor A - Commercial	450	-	450	=	1.00
Vendor B - Commercial	415	-	450	=	0.92
Vendor C - Commercial	400	-	450	=	0.89

Price Evaluation

The third area is the total price, evaluated and normalized on a similar type of weight factor.

Vendor A—Lowest price per unit	$90,000
Vendor B—Second price per unit	$93,000
Vendor C—Third price per unit	$96,000

Here, to normalize the lowest price to 1, you use the lowest price as a numerator and divide this constant numerator by each price.

Normalizing the price information

A = $90,000 is 1.00
B = $93,000 is 0.97
C = $96,000 is 0.94

The previous two areas are carried over for a final evaluation.

Normalizing the technical information

B = 455 is 1.00
A = 420 is 0.92
C = 400 is 0.88

Normalizing the commercial information

A = 450 is 1.00
B = 415 is 0.92
C = 400 is 0.89

Total Combined Matrix Analysis

Using a 40% consideration for price, 40% consideration for technical, and 20% for commercial, the following process takes place.

The final result is the percent balance times the normalized information.

		Score		%		Summary
Vendor A	Price	1.00	×	40%	=	0.400
	Technical	0.92	×	40%	=	0.368
	Commercial	1.00	×	20%	=	0.200
		A Total				0.968

		Score		%		Summary
Vendor B	Price	0.97	×	40%	=	0.388
	Technical	1.00	×	40%	=	0.400
	Commercial	0.92	×	20%	=	0.184
		B Total				0.906

		Score		%		Summary
Vendor C	Price	0.94	×	40%	=	0.376
	Technical	0.88	×	40%	=	0.352
	Commercial	0.89	×	20%	=	0.178
		C Total				0.906

Vehicle "B" is the best value, although it costs $3,000 more to purchase than Vehicle "A."

Vendor B is technically on spec but not the lowest price; it is not the best commercial risk but overall is the best potential quality unit. The percent balance of technical and price at 40% and commercial at 20% allows for small price variances to be normalized in technical specification quality and value.

The balance could be struck in favor of price percentage, but that reduces the technical integrity range. Technical integrity, being based on life cycle cost information, would impact purchase price, and this is the basis for life-cycle information to be collected and applied to the purchase effort.

The goal here is to provide a reasonable statistic evaluation—not to cover your decision process with meaningless data, but rather to employ an effective cost-evaluation process based on the profit dollar.

Maintenance Cost in Vehicle Bid Analysis

The following logic is an example of using historical maintenance information to further impact your analysis.

Vehicle "A" costs $10,000 more to maintain over 10 years than Vehicle "B." You can add the $10,000 to the price of Vehicle "A" or deduct it from the price of Vehicle "B," further reinforcing the efficiency of Vehicle "B." You should know the components cost areas that make up the $1,000 per year increased average cost validating your analysis.

People auditing can agree or disagree with your logic, researching your facts to their satisfaction. Your strategy is to be upfront with information prior to the award.

Variable Analysis Criteria

You can also use 50% price, 30% technical, and 20% commercial as a breakdown for various reasons. Your need is to agree on your priorities for considerations and implement them.

The following is the result of the same matrix bid analysis using the 50, 30, 20 format.

		Score		%		Summary
Vendor A	Price	1.00	×	50%	=	0.500
	Technical	0.92	×	30%	=	0.276
	Commercial	1.00	×	20%	=	0.200
			A Total			0.976

		Score		%		Summary
Vendor B	Price	0.97	×	50%	=	0.485
	Technical	1.00	×	30%	=	0.300
	Commercial	0.92	×	20%	=	0.184
			B Total			0.969

	Score		%		Summary
Vendor C Price	0.94	×	50%	=	0.470
Technical	0.88	×	30%	=	0.264
Commercial	0.89	×	20%	=	0.178
C Total					0.912

Best Value - #1
> Lower Bid
>> Second best technical

Communicating with the Vendor

You should share with the vendors your analysis format, reviewing the areas of focus, continually working toward objective information to support this process. The nature of this process tends to be subjective.

Visitation Process in Commercial Evaluation

The commercial area can become more objective by your visiting the vendor and quantifying its operation with a numerical evaluation.

Some general areas of focus are:

Factors for Evaluating Vendors/Carriers

Financial	Management	Technical/ Strategic	Relational
Income statement	One-man show, or	Quality R&D-	Sense of ease in
Balance sheet	balanced?	capability	adapting to our firm
Cash flow	Succession of management	Efficiency of	Quality commitment
Credit stature	Control mechanisms used	process/service	to us
Sales History	Culture or view of	Industry leader or	Customer service
	customers	follower?	elements
	What is the ownership?	Quality systems	EDI/order entry
	Warranty integrity	Supplier's suppliers	efficiency
		and laborers	Terms (FOB, sales,
		Destructive Testing	payment)
			Cost containment and
			sharing
			Transportation logistics
			factors

Quality Fundamentals

- Quality is management responsibility. It must start at the top and permeate the total organization.

- Quality improvement can come only from competent, committed people who are constantly searching for ways to do their jobs better.

- Quality improvement leads to increased productivity, market penetration, and profitability.

- Quality is primarily a state of mind. Improvement requires persistent attention to detail.

- Success requires a belief that everyone in the organization has value and a potential for creativity and contribution beyond his or her job description.

- Quality is free, but it is not a gift.

The Changing Image of Quality

Old Image:

- Low quality is caused by the worker.
- Some defect level is acceptable.
- Focus on the product, catch mistakes, and fix them.
- Buy from the lowest bidder.
- Quality is secondary to profits.
- Quality assurance is responsible for quality.

New Image:

- Poor management is the cause of most quality problems.
- Zero defects must be the goal.
- Focus on the process; prevent non-conformance.
- Buy for total value, which includes quality and reliability.
- Initial investment in quality is the key to long-term profits.
- Quality is the responsibility of the performer of the task.

Figure 10-1
The Quality Assurance System

Why Do Quality Efforts Fail?

1. Vague or inaccurate perceptions of what quality means. If quality is perceived as merely meeting technical specifications or doing good repair work, only mediocre results will be achieved. Quality must be perceived as managing for excellence in all operations.

2. A lack of solid commitment and backing by top management. Low or wavering commitment of energy and support will scuttle even the best-conceived improvement effort. This condition often stems from a poor understanding of the role of quality in all corporate functions.

3. Treating quality improvement as only another temporary program with a lot of slogans and banners. Top management's lack of commitment is obvious in this situation. Department managers and supervisors will view it as another onerous task to add to their already full plate. Token participation is all that is evoked.

4. No acknowledgment of the required cultural change. As Kaoru Ishikawa said, total quality control represents "A thought revolution in management. It is management...based on respect for humanity."[3] As such, it demands new ways of thinking and working. People will naturally resist changing their perception of what their jobs are.

5. Executives' belief that a high level of quality can be achieved merely with new techniques and devices, a few statistical charts, etc. The truth is that although advanced methods and process control technology are important for quality, they alone will not suffice.

Manufacturer/Supplier Evaluation

1. Financial Analysis
    a) Should be performed by Purchasing Group
    b) Should be performed annually
        (1) A must if dealing with a new vendor
        (2) Purchasing should maintain a list of accepted vendors
    c) Current ratio = Total current assets ÷ total current liabilities 2:1
    d) Debit to worth ratio = Total liabilities ÷ tangible net worth
        (1) Dun and Bradstreet reports
    e) A private firm or public
        (1) Financial information should be held in confidence
        (2) Should be willing to submit necessary information
    f) Study should be both short and long term

2. Human Resources
    a) Start with plant or facility visit
        (1) Observe cleanliness
        (2) Observe equipment—new, upgraded, or outdated
        (3) Observe worker attitude: positive, happy, secure
    b) Management
        (1) Is there a logical organization?
            (a) Is there an organization chart with responsibility assigned?
            (b) Are they knowledgeable in the field?

---

[3] Kaoru Ishikawa, *What is Total Quality Control? The Japanese Way*, (Englewood Cliffs, NJ: Prentice-Hall, Inc., 1985).

        (c) Do they have both short-range and long-range plans?
        (d) Are they sincere, and is there integrity?
        (e) Are they cost conscious?
        (f) Is quality control an important factor in the organization?
        (g) Does the organization have a good reputation and history?

  c) Work Force
     (1) Are they well trained?
     (2) Do they have certification where necessary?
     (3) Is it adequate numerically?
     (4) Are the employees concerned, or merely going through the motions?
     (5) Is there management/employee communication?
     (6) Are employees neat and clean?
     (7) Are employees experienced, and do they have time in the organization?

3. Engineering
  a) Is it the state of the art?
  b) Is it continuing or only new product?
  c) Is it practical as well as technical?
  d) Does it support service activities?
  e) Are experience and theory available?
  f) Are individuals qualified (professional engineers, etc.)?

4. Training
  a) Is operator training provided?
  b) Is maintenance training provided?
     (1) Are manuals up to date and specific or general?
     (2) Are items not in the manual documented by drawings or procedures?
     (3) Are instructors experienced technically and practically?

5. Impression Factor (Gut Feeling)
  a) Is there experience?
  b) Are there associated supply organizations?
  c) Is there integrity?
  d) Do they commit time?
  e) Is their reputation developed?
  f) Do they accept user input?

g) Do they participate in the industry in general?
h) Do they have an order backlog?

Product Evaluation

1. Design (New)
   a) Review of engineering criteria (is it innovative, well thought out?)
   b) Is there an ability or willingness to meet user needs?
   c) Has safety been incorporated into design?
   d) Quality
   e) Is the design applicable to job task?
   f) Other early user experience
   g) Serviceability
   h) Was the equipment field tested in addition to lab or plant test?

   Design (Established)
   a) User experience—in-house or other customers
   b) User acceptance—in-house or other customers
   c) Engineering support (modifications or improvements)
   d) Has safety been incorporated into the design?
   e) Serviceability of equipment and components
   f) Is the design stable or constantly changing?
   g) Past equipment failure reports

2. Engineering
   a) Safety factors—for all materials—both ductile and brittle
   b) Quality control involvement
   c) Staff qualification—Professional Engineer (P.E.) certifications
   d) Previous experience/knowledge
   e) Problems—responsiveness

3. Used Equipment Market Value
   a) Ways to establish
      (1) Auction—internal—on site or at a public
      (2) Bid lists—qualified bidders—sealed bids
      (3) Trade-in
      (4) Equipment journals or trade papers

4. Other Factors
    a) Parts pricing—excessive or reasonable
    b) Parts availability—on hand—short time—long delay

Vendor Visitation Rating

You should quantify each area with a 0–5 rating, similar to the ratings used on the technical and commercial areas, and share this information with the vendor. Periodically revisit and evaluate changes. Use this information to support the commercial evaluation summary.

Transportation should be practical, logical, and sensible in quantifying elements of the bid evaluation from each vendor. The vendor should know its strengths and shortfalls evaluated by your company to strive to improve.

Transportation should be as objective as possible to ensure your company is obtaining the best value for its money.

Vehicle Bid Analysis Summary

The elements of this process are continually changing. You must monitor these changes objectively and be flexible in your process to maximize your values.

Chapter II

Awarding a Bid

The Process of Awarding a Bid

You know the vendor to whom you'd like to award the contract for building a vehicle for your needs. You have evaluated the functional bid in terms of technical, commercial, and price, evolving the best vendor through a matrix format.

You will call the vendor, prior to notifying him and the other bidders of your decision, to discuss any discrepancies. The purpose here is to ensure that the vehicle on which the vendor bid was what he understood that you wanted.

Also, you asked for a line item for line item quote. You might want to remove some line items, lowering the price. You should look at historical data and past warranty claims specifically that were not paid by the manufacturer to be applied as credit toward the purchase price, further lowering the cost of the unit to you.

Before a sale is finalized, the bidder is very flexible, and his desire to close a sale opens an opportunity for you to bring up your past difficulties in collecting warranty claims. You might fare better if you brought up at this time these past claims turned down for a second consideration. Most warranty claims can be settled in conversation. Many times, written procedures are initiated by the fleet at the manufacturer's request for the manufacturer to evaluate the fleet's problem. The cost of paperwork in most instances outweighs the preparation of the warranty claim discouraging your efforts. When all these claims not awarded are banded together, they equal a cost worth bringing up and negotiating at a time when leverage is on your side. Immediately before a bid is awarded is a time when this process is in your favor.

263

Negotiating

Negotiating is more of an art than a science. It is most successful when both the seller and buyer become winners. You, the customer, have expectations of an acceptance level, and the vendor miust come close to it through this bid. Each element of the bid was priced. The vendor will not give his cost to put together the product. The vendor's cost is composed of:

- Direct costs
- Administrative costs
- Profit

Warranty claims will impact the vendor's profit and administrative cost areas. You should cite the vendor's administrative time in processing your warranty claims as a savings on the vendor's part by deducting these claims from the price of the vehicle or as a credit in ordering parts and services from the vendor.

A good supply of repair orders as documented costs help to support and explain the nature of your request. You can review the labor and parts, item for item, or classify them. Examples should be: "average cost of $500 per vehicle for the last 25 vehicles you delivered to us . . ." or ". . . $100 in electrical costs, $200 in engine costs, $100 in suspension costs, and $100 in frame costs. . . ."

Most of the potential problems are with the mounted equipment vendor that assembles the body and mounted equipment on your chassis. The manufac-turers of the chassis and mounted equipment are required to adhere to many industry standards. The body manufacturer is more flexible and does not have as strict a standard to which to adhere as the chassis and mounted equipment manufacturer. For the body, the customer should quote thickness and mixture of metals and body preparation for paint, which determine longevity of product in its application.

The clear delineation of past day-to-day problems and the preventatives of these items in a simple, logical, understandable, and creditable format is most important to utilize leverage. In effect, you are creating an environment to allow a win-win situation to evolve naturally and easily. You define points of disagreement. You define misunderstandings created in the bidding process. Identify perceptions that are not valid to clear questions that were not seen in the bid package by both sides as they should have been seen.

An example of clearing the air of misunderstandings is two people arguing over who gets the last orange. Both aggressively want the orange and will not relent. The negotiator asks, "What do you want the orange for?" One wants only the flesh to eat, and the other wants the skin to shed for baking. When their needs are defined, they each agree that rather than splitting the orange and having 50% satisfaction, each can be 100% satisfied by working together. A win-win situation.

Time Frames

You must define the ordering process, acceptance of chassis and body, delivery to the mounted equipment dealer, and assembly. Dates should be established, penalties agreed upon, and payment terms set. Definition of inspection timetables and at whose expense should be clarified for pre-build, pre-paint inspection, and final inspection. Timetables must be agreed for scheduling these inspections and delineating at whose cost to perform these visitations and, should additional inspections be necessary, at whose cost.

It is better to inspect the first unit at pre-paint to allow adjustments to take place. The manufacturer should have the staff to perform this task, using your functional specification and drawings for a reference point, thus minimizing the corrective work needed for acceptance. If the customer must perform this function, credit should be assigned to the customer. It is the manufacturer's and the dealer's responsibility to prepare a "turnkey" product to be delivered to the fleet to be put into service.

Testing

You should verify what sampling is to be used to check the quality of the major and minor components and destructive testing as needed. Critical areas of concern should be outlined, and the manufacturer should define the testing needed and the frequency. An example is every "hydraulic lift cylinder" will be x-rayed and inspected to comply with American Welding Society standards with appropriate documentation. You should inspect this process to ensure it is taking place. Support documentation must be identified for reference, for example, destructive testing as outlined in Figure 11-1.

Pre-Build Meeting

The last time you purchased a vehicle from this successful bidder, you had some problems. You should list these problems and review what corrective

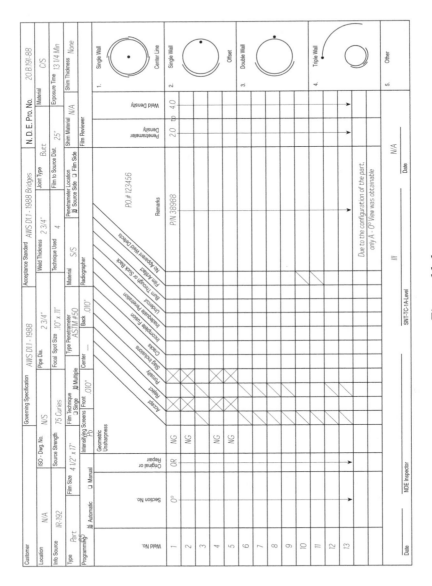

Figure 11-1
Destructive Testing Validation

action was taken last time to fix the problems and what is in place now to preclude a repetition. If nothing is in place, what can you put in place to facilitate a turnkey delivery? Figure 11-2 shows an example of what would be listed.

Chassis
1. Change cab to axle to 126 in.
2. Transmission will be a Fuller FS 7206—6-speed.
3. Rear axle will be RT40-145P with controlled traction.
4. Air dryer to be Bendix AD9.
5. Frame not to be dropped. Use straight frame.
6. Auxiliary steps on fuel tanks to be 1-1/4 in. cable sides with non-skid tread area. Step to protrude out past next step above (staircase effect).
7. Delete battery box location 22 in. from cab. Stay with standard location of battery box.
8. Front axle to include driver-controlled traction differential.
9. Check on availability of skid plates.

Body/Boom
1. Delete 9-in. riser under tower because chassis frame height on 6 × 6 unit will be 8 in. higher than normal chassis.
2. Outrigger mountings to be adjusted for proper ground penetration due to higher frame height.
3. Locking hand throttle may be supplied with chassis. Credit may be due if supplied.
4. Delete HP 2200 Fairmont double-acting valve and HP 2201 10-ft. hose assembly. Retain high-pressure output fitting and return fitting on Fairmont press. Cutout in control cover to remain so hose assembly can be coupled to press.
5. Bucket pad supports to be 24 in. long × 2 in. wide.
6. Basket lip liners to be pop riveted to liner. Use nylon rivets.

Figure 11-2
Pre-Build Meeting Items for Concern

After all these activities are complete, you make the award, sign a contract, and validate the delivery date and inspection dates.

Chapter 12

Vehicle Delivery and Warranty
✚

When the assembled vehicle is ready for delivery, several areas should be addressed. When the chassis is completed at the manufacturer's facility, it should be inspected to verify any spec changes prior to delivery to the mounted equipment dealer, especially if it is the first of many. Not only does it validate the manufacturer's quality control, but it also offers an opportunity to those who participated in the vehicle specifications a chance to realize its final form. Serious manufacturer-related problems also can be addressed prior to mass-production errors.

When the chassis is delivered to the body and equipment dealer, a review of a prototype of a large order or inspection of the one or two units produced is necessary to confirm all the widgets are in the right place per spec. Only after this process is exercised will you start the process to pay for the unit—including the chassis, the body, and the mounted equipment. This must be detailed and included in the specification so when the vendors bid, they know the terms of payment and performance requirements.

Also, most fleets have automated management information systems, and the input of spec information into your management information system should be of primary concern. One input document that is a point of reference is the vehicle and equipment birth certificate. It should be included in your specifications and requires the manufacturer to identify this information in your preferred format, saving your searching for the information on multiple documents and on the vehicle before you enter it into your information system.

Specifications are a clear statement of needs. The current timing of this request is prior to the specifications being awarded to the successful bidder. This engagement, or romantic period, is the only time to bring up details that lead to a happy marriage:

- Warranty recovery parameters
- Performance and standard details
- Delivery and payment details
- Technical components
- Timetables

The first year you have the vehicles in service will be unpredictable, so the scene must be set prior to the acceptance of the bid for the proper disposition of these surprises to the mutual satisfaction of the chassis, body, and equipment manufacturers and the user. This is better done before the deal is finalized, lest there be disappointments.

Life-cycle costing with maintenance and operating costs will identify a reasonable time to economically replace a unit. You should have similar units available so that in a timely manner you can replace vehicles as needed rather than all at once. Space orders so they arrive 10% to 20% per month, and be flexible to replace the worst vehicles first. This cycling allows a better distribution into your fleet and covers the unexpected, premature vehicle replacements due to accidents, etc., which are always present.

This kind of data can be handled manually. However, in a large fleet, some automation is necessary because of the accuracy that is needed for clean and credible information per your automated management information system needs.

Why extend this effort? Each fleet has an operating application that a class of vehicles serves. This climate is responsive to geography, people, cargo, and management style, and each class of vehicle will have an average economic life. You should fix this life cycle and generally stick to it for a cost-effective operation. As the economic climate changes (interest, profitability) and your maintenance productivity changes due to people mix and technical changes, you can measure these factors and initiate action. The main purpose of this effort is for you to be an efficient part of the total management of the company annual budget process. You can set a vehicle replacement policy that can be budgeted for efficiency. This policy should be one that, if vehicle replacement cycles are extended or shortened, their cost change in your operating and maintenance costs can be adjusted realistically (domino effect) so that company management is not surprised by increased manpower or finance costs. The goal here is to plan your work and work your plan. A team can better accomplish this goal with realistic statistical support; in effect, you should be accurate fortunetellers based on historic patterns.

Once a year, review the fleet age and cost, determine vehicle replacements due to life-cycle maintenance and operations cost information, and order for timed delivery new units to replace old to provide flexibility to your changing conditions.

The more complex the analysis, the more room for error. This methodical approach lends itself to a controlled environment; fine-tuned annually, it supports a profitable operation.

When the vehicle chassis is complete, it is delivered to the mounted equipment dealer to put the body and mounted equipment on the chassis of the vehicle. At the point, where the assembly is put together before painting, the customer should inspect the first unit using the vehicle specification as a guide. The unit should comply to your specification.

A list of problems should be generated for the manufacturer or dealer to correct. Copies should be sent to the receiving garage for its in-service inspection process. An example is:

Vehicle 1234

1. R × R outside tire flat
2. Doors on 18 × 20 bin do not close
3. Outrigger rear street side leaking
4. Sub-frame vertical welds over right rear wheel not acceptable

These problems should be corrected and the vehicle painted and shipped to the fleet for acceptance. Upon completion of the corrective action, the receiving garage should be notified of the delivery date.

When the vehicle is delivered, it should be signed for as being delivered and noted that delivery does not constitute acceptance.

Within one week, the unit should be inspected using the pre-paint letter as a guide to verify the corrective actions initiated by the manufacturer, with the vehicle specification as a guide. In addition, the in-service inspection by the company should be performed. An example is shown here.

Sample In-Service Inspection Process for a Tractor

Spending time on a new vehicle in-service inspection can offer trouble-free service, extend vehicle life, reduce vehicle down-time, and pay for itself many times in maintenance costs savings, fuel savings, and provision for a sound basis for a warranty recovery program.

A. **Power Train**—Because this is the most expensive system to repair, start with the engine. The new electronic engines have a diagnostic capability. Is the engine performing at rated standards? Check the oil level, and know the type of oil to be used. An initial oil change and oil filter may be desired. Check the system for leaks.

B. **Cooling System**—Pressure check the system, note leaks, and check the antifreeze for proper concentration. Check belts, hoses, and clamps for routing and proper support.

C. **Fuel System**—Drain the primary fuel filter and tank to remove sediment and to ensure there is no water contamination. Check connections and fuel lines.

D. **Start-Charge System**—Load test batteries, and check at various locations for voltage drop. Ensure that connections are clean and tight, cables are of proper size, wiring is properly clipped and routed, and alternators and starter bracket fasteners are properly torqued.

E. **Brakes**—Brakes must be in balance, and slack adjusters installed at proper angles. Check lining to drum clearance, push rod travel, and lining thickness. Check for air leaks.

F. **Lighting**—All connections should be clean and tight.

G. **Tires, Wheels, and Alignment**—Check for proper air pressure. Check that all wheels have proper oil level and no leaks. Check that all lugs are tightened. Check for sufficient clearance at dual wheels and toe-in adjustment.

H. **Suspension and Steering**—Check steering for excessive play, and tighten U-bolts.

I. **Cab and Instruments**—All gauges must be working properly. Air conditioner should be checked for temperature output.

J. **Add-Ons**—Accessories installed at the fleet level can save substantially on the initial purchase price.

K. **Supporting Papers that Accompany the Vehicle**—These include manuals, vehicle spec sheets, and wiring diagrams.

L. **Fifth-Wheel Mounting (if Tractor)**—Slide should be in working order, and fasteners should be torqued.

With 15,000 parts on the vehicle, if the manufacturer was 99% efficient, that leaves 150 items of concern. You should operate the unit to verify that all components work. Prepare to put the vehicle in service to shake it out for 30 days.

At the end of the 30 days, if all is well, initiate the payment process. If any problems occur during the 30 days, require the manufacturer to correct them. If not corrected by the manufacturer in a reasonable time (e.g., 60 days), hold payment and possibly reduce the price of the vehicle due to extra maintenance and operating costs incurred to the old vehicle during in-service delay. Complex vehicles require complex acceptance terms and a strategy to apply leverage. When you pay for the vehicle, that vehicle should be what you expected.

If that vehicle is not what you jointly expected, then you must bring it up to joint expectations or reduce the price of the vehicle as outlined in your liquidated damage requirements.

The key personnel here in the acceptance process are the user and the shop. You give the user training on the correct operation of the unit. You train the shop in preventive and diagnostic maintenance requirements of the unit.

To accept this vehicle, both the user and the shop must state that the vehicle is acceptable. You require the user to turn in his old vehicle and the shop to accept the old vehicle and authorize payment (Figure 12-1).

_____ **GARAGE**

FINAL ACCEPTANCE

Vehicle Number _____

Chassis Manufacturer _____

Chassis Dealer _____

Chassis Serial Number _____

Equipment Manufacturer _____

Equipment Dealer _____

Equipment Serial Number _____

Date Delivered to Garage _____

Date Inspection(s) Completed _____

Chassis Acceptable _____

Chassis Not Acceptable _____

Equipment Acceptable _____

Equipment Not Acceptable _____

Date Prepared _____

Prepared by _____

Authorized by _____

Figure 12-1
New Vehicle Acceptance

Warranty

Warranty is a procedure whereby you attempt to recover in part or total repair costs incurred by your company from original equipment manufacturers, suppliers, or rebuilders every time it is determined that an item has failed due to faulty workmanship, material, or construction.

A vendor expresses his confidence and knowledge of the product in the warranty terms he extends.

As a buyer, you should evaluate the manufacturer's published warranty as a statement of confidence in the vehicle being built to specifications and being operated satisfactorily with reasonable care and minimal repair requirements. It should provide the buyer with protection against high operating costs due to premature parts failure or poor assembly workmanship.

Suppliers generally offer the same vehicle and component warranties to anyone who purchases their products. However, for the fleets—depending on philosophy, programs, procedures, and size—warranties can become specific tailored contracts.

Warranty Categories

The following lists the primary types of warranties:

- Standard
- Optional
    Extended
    Defects quantified and levels set
    Policy adjustments
- Extended component
- Component
- Vehicle
- Customized
- Chassis
- Mounted equipment

Warranty may be summarized and placed in four different general categories:

1. Standard Warranty

    A standard warranty is an offer extended by the original equipment manufacturers to cover faulty workmanship, material, design, or construction of the product. This warranty carries a specific time limit, hour limit, or mileage, whichever occurs first.

    Basic standard warranties are:

    - Powertrain coverage
    - Major component coverage
    - Corrosion coverage

- Emissions control system
  Emissions Defect Warranty
  Emissions Performance Warranty

2. Extended Warranty

   An extended warranty is a mutual and/or purchased agreement between the manufacturer and you that extends the basic standard warranty policy in either time, mileage, or both.

   Types of extended warranty contracts would be:

   - Extended service plans
   - Special service vehicle optimized contracts
   - Corrosion service contracts
   - Environmental protection plans
   - Fleet service plans

3. Policy Warranty

   A policy warranty is a mutual agreement between you and the manufacturer/supplier whereby adjustments are made to cover chronic failures, latent defects, and problems that occur over and above the basic standard or extended warranty coverage.

   Policy warranty programs include:

   - Owner notification programs
   - Good-will adjustments
   - Service recall programs

4. Specific Service Warranty

   These types of warranties are seldom published to the public and involve unique trade situations. They are contractually established on a one-to-one basis by OEMs. Full-service lease fleets have successfully negotiated 24-month/24,000-mile specific service warranty agreements with the OEMs and body fabricators.

The following are sample terms and condition wordings to attach to specifications as a "boiler plate" for reference to claims and/or understandings or expectations of the company.

Standard Equipment

The unit proposed must be equipped with the manufacturer's equipment and accessories, which are included as standard in the advertised and published literature for that unit. No such item shall be removed because it was not specified. To promote fairness in competition, items offered as standard equipment by a particular manufacturer, but not recognized as such by the industry, should be clearly noted. An option to delete those items at an offered price may be included with the bid package. All pins and fasteners used in the construction of this unit shall be graded and certified as being genuine, shall be engineered as correct in their application, and shall comply with "best manufacturing practice" standards. All critical pins and fasteners must have evidence of destructive testing to ensure minimum standards have been reached.

Silence of Specifications

The apparent silence of this specification as to any details or the omission from it of a detailed description concerning any point shall be regarded as meaning that only materials of first quality and correct type, size, and design are to be used. All workmanship is to be first quality. The unit herein specified shall be constructed throughout of new parts and materials, which shall not have been serviced other than that necessary for the factory tests, unless otherwise allowed by this specification. The unit bid must be the latest model offered. All interpretations of this specification shall be made on the basis of this statement.

Compliance with Regulations and Standards

The parts and procedures employed in mounting, affixing, plumbing, wiring, painting, and finishing each unit must ensure compliance with applicable regulations as standards. Additionally, each completed unit must conform to regulations and standards as established by federal and state regulatory agencies, Occupational Safety and Health Administration (OSHA), Society of Automotive Engineers (SAE), American National Standards Institute (ANSI), American Welding Society (AWS), ASTM International, Federal Motor Vehicle Safety Standards (FMVSS), Institute of Electrical and Electronics

Engineers (IEEE), and other agencies. All welding and weld inspection must be in accordance with applicable AWS standards. Each unit must be certified to comply with applicable standards.

Uniformity

Multiple units ordered pursuant to a single specification must be identical in components and operation, except as provided within the specifications. All structural, wiring, and hydraulic diagrams must reflect this commonality.

Compliance with Specifications

Any exception to this published specification must be specifically noted in both the priced and unpriced responses to this request for quotation. Alternatives may be bid, but it usually will be in the best interest of the respondent to initially offer as close to specification as possible, followed by proposed alternatives and their effects on pricing.

It is requested that in the technical response, the bidder responds to each specification item. If the bidder complies, a check mark, a "yes," or other symbol may be used to so indicate. Exception should be fully disclosed and explained.

Descriptive literature displaying the equipment proposed is to be provided. A demonstration of the equipment model may be required prior to award to demonstrate that the unit is in production and meets or exceeds requirements.

Inspections

The company personnel must be given reasonable access to monitor the production of the bid for the unit. Typical meetings, although not required in all cases, would include the following:

- Pre-bid meeting to discuss the specification details and specific line items

- Pre-award meeting to verify that the proposal has been prepared in a manner that will satisfy the company specification intent

- Pre-construction meeting to review the design and blueprints for the unit

- Pre-paint inspection of the unit

- Pre-delivery inspection of the completed unit

Progress inspections also may be conducted at various times to observe critical operations. Final testing prior to delivery may be performed during one of those times.

Delivery

The successful bidder will deliver the completed unit or units as instructed to the company garage in each city. At delivery, drawings, manuals, applicable testing certifications, and other documentation must be presented. Wiring diagrams, hydraulic schematics, and flow diagrams will be required. A comprehensive parts manual, including exploded component diagrams, parts numbers, and any parts crossover numbers, must be accompanied by opera-tors' manuals and recommended preventive maintenance program tailored to the unit or units received. Weight slips indicating total equipment weight also must be provided. Additional weights for a bare cab/chassis also will be required. Delivery does not mean acceptance for payment nor does it consti-tute acceptance.

In-Service Demonstration and Instruction

Prior to field usage of any completed unit, a maintenance and in-service equipment demonstration and instruction session must be held to familiarize the users with the new equipment. An explanation of the features and capabilities of the unit should be tempered with cautions of the limitations of the equipment. The unit should be made available for monitored instructional operation as a part of that session. A separate session with the system mechanics of the company must address the recommended lubrication, inspection, and problem diagnosis for maintenance of the unit, which includes any special diagnostic tools necessary to perform these tasks. Persons per-forming such in-service training must be qualified in that capacity and approved by the company in advance of the demonstration.

Sample Boiler Plate

Buyback:

Vendor to offer a guaranteed repurchase figure, at which it will buy back the proposed piece of equipment at the end of three (3) years or 5000 mi (whichever comes first). The company reserves the right to exercise this option.

Buyback Figure $_____

Service:	**Yes**	**No**
1. Vendor to supply one complete set of track loader, application-specific parts, operator, safety, and service manuals, hard copy, and CD-ROM.	____	____
2. Vendor to supply a complete first 250-hour service kit.	____	____
3. Vendor to have complete parts and full-service facility located within an accessible radius of the company.	____	____
4. A parts line sheet will be supplied by the successful bidder that is directly referenced to the accepted equipment.	____	____
5. Any substitutions will be subject to the approval of the company director of fleet management.	____	____
6. Provide a list of parts needed for one year of maintenance.	____	____
7. Provide a detailed, sequenced PM program for equipment maintenance.	____	____
8. All parts must be available for delivery within 24 hours.	____	____
9. All warranty claims must be addressed within 24 hours.	____	____
10. End Product Questionnaire (EPQ) must be addressed within two weeks of delivery.	____	____

11. A list of critical parts destructive testing must be provided to the company prior to delivery.

12. An 8-hour training session must be provided by the manufacturer to three equipment operators and three mechanics prior to final acceptance. This session shall include operating features and benefits, daily inspector PM check points, and loading and unloading procedures for transporting to and from job sites.

13. The Contractor acknowledges that he is an expert fully competent in all phases of the services required under this contract. The Contractor warrants that he shall, in good workmanlike manner, perform all work and furnish all supplies and machinery, equipment, facilities, and means, except as herein otherwise expressly specified, necessary or proper to facilities and means, except as herein otherwise expressly specified, necessary or proper to perform and complete all the work required by the contract. Although the Owner may review and approve various documents including, but not limited to, quality assurance manuals, installation agreement, and so forth, the Owner neither accepts in any manner responsibility for nor relieves the Contractor from responsibility for the performance of all the requirements of this contract. The Contractor warrants that all material furnished will be free of any defects in material and workmanship, and will be the kind and quality as specified by the purchaser. If any failure to meet the foregoing warranty appears within one (1) year after the date first placed in use, the Contractor will correct by repair or replacement and provide that each such repair or replacement will carry a two-year (2) warranty from completion of said repairs or replacement. The Contractor shall be responsible for all removal costs into and out of the location(s) the Owner has designated wherever a repair or replacement is required.

    The foregoing shall not negate any other warranties of the Contractor, expressed or implied by operation of law.

14. Article—Latent Defects

    Notwithstanding any other clause in this contract, should a "Latent Defect" in the work be discovered during the five (5) year period after the work has been accepted by the Owner, it shall be the Contractor's responsibility to repair or replace the work. A "Latent Defect" is defined as a defect that reasonably careful inspection will not reveal, or a design defect that could not have been discovered by inspectors, or a defect that could not be located by any known or customary test.

281

15. Liquidated Damages

If a unit is not fully accepted when delivered at the company garage and such delay is the responsibility of dealer or body manufacturer, all costs for a compatible vehicle rental will be paid by the Contractor. If the vehicle is not turnkey at the end of 60 days from acceptance at the company garage, then the vehicle will be returned at no cost to the company and total monies returned to the company within 14 days of return.

An End Product Questionnaire (EPQ) must be supplied within two weeks of delivery.

Four categories of warranty are available from four separate sources:

1. Original equipment manufacturers
2. Component manufacturers
3. Parts replacement suppliers
4. Rebuilders

Warranty failure analysis should follow these guidelines (see Figure 12-2):

- State the problem clearly and precisely
- Organize logically and record the facts
- Identify probable cause
- Communicate with the staff responsible for the cause
- Make proper repairs
- Follow up to ensure that the problem is fixed

Facts are available from three primary sources:

- Your jobsite
- The failed product
- Metallurgical labs

The root cause is the specific condition that started the problem. When the root cause has been identified, it is necessary to find the party responsible for the problem and take the necessary steps to prevent its recurrence when possible.

Unit No. _____ Make _____ Model _____ Year _____ Serial No. _____

Current hour meter_____ Hours since last service_____ Type Svc. _____

Hours since last reported problem_____ What _____

Action taken _____

Last oil analysis_____ Findings _____

File review _____ Yes_____ No comments _____

Current failure location or problem area:

Air intake _____	Transmission _____	Rear Axle _____	Hydraulics _____
Electrical _____	Undercarriage _____	Tires _____	Suspension _____
Attachment _____	Structure _____	Cooling Sys _____	Turbo _____
Fuel _____	Lubricant _____	Exhaust_____	Engine _____
Block _____	Head _____	Valves _____	Cam _____
Pistons/liner _____	Rods _____	Crankshaft _____	Bearings _____

Other: _____

Failure (statement of current problem) _____

_____

_____

_____

Operator's comments at time of failure: Where operating: _____

How: _____

Immediately before failure: _____

_____

Immediately following failure: _____

_____

Mechanic's observations: _____

_____

Figure 12-2
Sample Failure Report

There are steps that must be followed for proper failure analysis. The objective of failure analysis is to identify the correct root cause. Doing the steps out of sequence or skipping steps might not lead you to the root cause.

1. State the problem clearly and concisely

2. Organize fact gathering

3. Observe and record the facts

4. Think logically with the facts

5. Identify the most probable root cause

6. Communicate with the party responsible for the failure

7. Make repairs as directed by the responsible party

8. Make sure the problem is fixed and steps are taken to prevent a repeat of the problem

Information can be:

1. False—Always wrong

2. Assumptions—Often wrong

3. Opinions—Depends on the credibility of the source

4. Feelings—Often lead to exaggerations and errors

5. Facts—The actual occurrences

Always:

1. Insist on getting the facts, especially in important areas

2. Keep asking, "Am I getting and recording the facts?"

3. Ask quantitative and qualitative questions to ensure you are getting the facts

Warranty Recovery Package

Items that must be addressed in the upcoming warranty recovery package are as follows:

A. **Warranty Information and General Guidelines**—Summarizes the various warranties with guidelines

B. **Reimbursement**—Provides information as to the terms of payment

C. **Warranty Repair Order Claim**—Outlines claim processing and guidelines, manual or automated

D. **New Vehicle Preparation**—Guidelines on received and inspection of new vehicles

E. **In-Transit Damage**—Guidelines for inspection and reporting entranced damage and loss

F. **Service Recalls**—Procedure to inspect and correct safety or emission-related problems in specified vehicles

Warranty can be recovered in a cash settlement or as service performed by the manufacturer. You target 2–5% of the vehicle purchase price as your goal for warranty recovery on each vehicle purchased, whether in cash or service rendered by the manufacturer after delivery within a declared warranty program.

An example of a fleet's warranty problems is summarized in Figure 12-3.

Note that in reviewing the high number of claims, the engine is the highest frequency, followed by frame, body, air intake, suspension, brakes, and lights. In terms of dollars, the engine is first, followed by frame, brakes, body, transmission and clutch, cooling systems, air systems, and suspensions.

Looking at cost per warranty claim, transmission and clutch is $425 per claim. Brakes follows at $350 per claim, and frame at $250 per claim. Body and cooling system are next at $200 each per claim, and engine and suspension are $130 per claim.

You should examine all problems, prioritize them, and do something. If you are not looking, you are enjoying a false sense of security and allowing costs to deteriorate your profitability.

WARRANTY STUDY

	0-25,000 mi		25,000-50,000 mi		50,000-75,000 mi		75,000-100,000 mi		Total	
	No. of Claims	Cost	No. of Claims	Cost	No. of Claims	Cost	No. of Claims	Cost	No. of Claims	Cost
Engine	54	6,081	63	2,476	19	3,699	7	3,138	116	15,394
Electrical	11	603	3	239	2	641	3	435	19	1,918
Trans/Clutch	8	4,904	2	265	2	112	2	520	14	5,801
Steering	5	115	4	1,152	2	105	3	190	14	1,562
Cooling System	7	266	10	712	3	587	5	3,502	25	5,067
Air Conditioning	9	404	4	1,099	2	669	3	205	18	2,377
Diff/Dr.Shaft	1	42	1	96	1	367	1	170	4	675
Fuel System	12	1,133	2	174	3	491	3	500	20	2,298
Wheels	9	777	2	195	0	0	2	149	13	1,121
Brakes	15	1,273	0	0	10	4,621	3	3,910	28	9,804
Air System	24	2,143	5	460	5	852	0	0	34	3,455
Bogie/Suspension	14	1,129	6	1,203	11	1,122	3	1,033	34	4,487
Hubodometer	31	976	1	60	0	0	0	0	32	1,036
Body	30	2,842	7	964	4	4,987	3	154	44	8,947
Frame	42	9,970	5	2,009	0	0	0	0	47	11,979
Fifth Wheel	8	272	3	138	0	0	0	0	11	410
Lights/Wiring	24	666	1	85	11	25	1	23	27	799
TSI/Everfill	14	1,269	3	126	1	30	0	0	18	1,425
Tachographs	6	0	1	0	1	0	1	0	9	0
Totals	324	34,865	96	11,453	67	18,308	40	13,929	527	78,555
Claims/Units	6.6	712	2.0	234	1.4	374	0.82	284	10.8	1,603
Unit Avg. Warranty	0.029		0.010		0.015		0.011		0.016	
CPM/Period										

Figure 12-3

Sample Fleet Warranty Summary (Courtesy Heavy Duty Trucking Magazine)

Sample Repair Campaign Messages

The manufacturers are receptive to your information so they can make a better product. You need accurate information to enter these conversations to benefit your bottom line. Your Vehicle Identification Number (VIN) can be used to dial up repair history campaigns related to your specific vehicle.

Special Service Message 2441:
Certain complaints of high steering efforts on F/B-Series vehicles equipped with Bendix steering gears are the result of pressure-relief valves, which are set at or below minimum specs. Many service personnel install a new steering gear assembly, which provides no more than nominal relief pressure. A pressure adjustment kit, Bendix #2264080, is now available through Bendix. The kit consists of three adjusting shims, a replacement sealing washer, and instructions. This message is continued on #2442.

Special Service Message 2442:
When applied to a high effort gear, it will be possible to adjust the pressure-relief valve to the upper limit of its spec, without removing the gear from the vehicle. Refer to the 1991 F/B-Series Shop Manual for relief pressure specifications; page 13-46-7 for F-Series with 10–12 K front axle and pages 13-48-26 and 2 for F/B-Series with 6–9 K front axle. This procedure will optimize the output of the steering gear only. Properly adjusting the steering gear PRV will make the system more sensitive to marginal pump pressure at idle. Continued on 2443.

Special Service Message 2443:
If the steering gear PRV is properly adjusted and the vehicle still exhibits inadequate power steering assist, pump performance and idle speed should be checked in accordance with the shop manual.

Special Service Message 2645:
Most of the power steering fluid leaks from around the threads of the valve nut surrounding the input shaft of the Bendix C300-N steering gear can be corrected with the procedures in TSB 89-19-15. If the steering gear still leaks fluid from around the valve nut after performing TSB 89-19-15, this would indicate that the threads of the valve nut fit tight enough in the housing that the Bendix sealant (Loctite 567) is wiped off the threads when the valve nut is screwed into the housing.

Conclusion

Purchasing a vehicle and/or equipment is not an off-the-shelf process. Each piece of equipment and/or vehicle has a job to do. It is a tool for productivity achievements. Work methods change due to newer tools, and we extend tool life through better design, technology, and experience from application-specific environments.

We seek knowledgeable associates to be up to date in what is available. Manufacturers of chassis bodies and mounted equipment, their dealers and distributors, our work crews, mechanics, and technicians together can partner to meet each other's needs efficiently through a functional specification and a cost-effective solicitation process. That is our mission of this text en route to this vision.

Index

Page numbers followed by an *f* or *t* indicate figures or tables, respectively.

OEMs
    maintenance costs reduction and, 135
    relationship with fleets, 131, 136
    role and responsibilities, 132
    training/information provision, 133–134
    warranty determination, 133
Operating costs. *See* Life-cycle costing
Operating expenses
    defined, 39, 40
    in total unit costs analysis, 74*f*, 76, 76*f*

Pre-bid conference, 157–158, 225–226
Pre-build meeting, 265, 267, 267*f*
Proposal forms, 234–235
Pusher axle and performance, 188

Quality assessment, 256–258

Radial tires, 187, 195–196
Radiator shutters and performance, 187
Rebuilding a vehicle, 29, 30
Reconditioning a vehicle, 30
Relifing/updating a vehicle, 29, 31
Remanufacturing a vehicle, 30
Replacement of a vehicle
    alternatives to, 29–31
    budgeting information needed, 52–54, 53*f*
    cost analysis, 29–31, 46
    decision guidelines, 46, 55–57
    decision revisited in bid analysis, 246–248
    economics of (*see* Economics of vehicle replacement)
    forecasting for, 28–29
    life-cycle costing and, 88–89, 89*f*, 90*f*
    policy setting, 2–4, 3–4*t*, 54–55
    priority-ranking approach, 57, 58–59*f*, 124
Reports
    independent condition reports for resale, 106, 109, 109*f*
    inspection reports needed for a resale, 106, 107–108*f*

About the Author

John E. Dolce is an active fleet manager, consultant, author, and industrial instructor with more than 35 years of experience in car, truck, transit, mining, and equipment maintenance and operations. His background includes public and private fleets ranging from 100 to 18,000 vehicles with 24-hour, 7-day multishop and multistate operating locations. Mr. Dolce has helped thousands of management individuals and companies in the public and private sectors to implement cost efficiencies nationally and internationally during the past 30 years.

Mr. Dolce earned an A.A.S. in auto and diesel technology from the State University of New York, Farmingdale, Long Island. He also earned a B.S. and an M.A. in industrial management, both from New York University, New York City. Since 1966, Mr. Dolce has worked as Operations Manager—Truck Training School for Rentar Educational Corporation in Long Island City, New York; Assistant Chief Engineer Fleet for Royal Globe Insurance Company in New York City; Director of Personnel, Safety, and Security for Wooster Express, Inc. in Windsor, Connecticut; Fleet Manager of 3,000 vehicles at the U.S. Postal Service in Newark, New Jersey; Fleet Manager of 5,500 vehicles for General Public Utilities in Parsippany, New Jersey; Vice President of Transit Maintenance for 400 transit buses for Liberty Lines, Inc. in Yonkers, New York; Vehicle and Facility Project Manager for Baker Engineering for five maintenance facilities, alternative-fuel equipped; Vice President and General Manager of Mail Contractors of America in Little Rock, Arkansas, for 1,800 vehicles; and Director Fleet Management Division for Essex County in Newark, New Jersey, for 1,000 vehicles.

In addition to his current role as an active fleet manager and a private consultant, Mr. Dolce teaches seminars in fleet and facility management nationally and internationally for George Washington University, the University of Washington at Seattle, and the Society of Automotive Engineers (SAE). He also teaches fleet management in the undergraduate and graduate programs at the City University and the State University of New York.

Mr. Dolce is a Trustee of the Equipment Managers Council of America, is active with the Association of Equipment Personnel and National Private Truck Council in their certification programs, and is on the Board of Directors of International Fleet Management, Inc. He is an SAE member and serves on various SAE committees. Other books authored by Mr. Dolce include *Analytical Fleet Maintenance Management* and *Fleet Management*.